Cilfái
Historical Geography on Kilvey Hill, Swansea

Cilfái

Historical Geography on Kilvey Hill, Swansea

Nigel A. Robins

Nyddfwch

Copyright © 2025 Nigel A. Robins

All rights reserved.
First Edition 2023
Second Edition 2025

British Library Cataloguing-in Publication Data
A catalogue record for this book is available from the British Library
No part of this book may be used or reproduced in any manner for the purpose of training artificial intelligence technologies or systems,

Nigel A. Robins
Nyddfwch Publishing
16 Voylart Road
Swansea
SA2 7UA
nyddfwch@gmail.com

ISBN: 978-1-7393533-0-8

Acknowledgements

This study has been heavily reliant on past teaching notes and lessons learned from students' questions and discussions over many years. I am indebted to all of them. Grateful thanks to the staff of West Glamorgan Archives Service and the archivists at the British Geological Survey in Keyworth who allowed me to spend time with William Logan's original notebooks. The help of experts such as Dr John Alban and Gerald Gabb has always been beyond value and they have helped me as a sounding board and unlocking further fields of expertise which ave been so valuable. The contributions and discussions with my oldest friend John Andrew on geology and the rocks of Townhill and Kilvey have been particularly inspiring.

I must also thank the staff and colleagues at Coed Cadw/Woodland Trust who, unwittingly perhaps, spurred me on to re-explore Kilvey some thirty years after I last surveyed the land in the late 1980s. All the modern mapping was completed using the open-source QGIS application which has become a central tool as a landscape historian over the past ten years. Finally, I must mention the help and support of Kilvey Woodland Volunteers. Without the passion and commitment of the volunteer body over many years, I doubt that Kilvey would be the special place it has become. As I write this, Kilvey is under more threats from the local council and developers, and I hope this little book records a few milestones in the ongoing Kilvey story rather than an ending.

The first edition of Cilfái was remarkably successful. The aim was to fill a gap in knowledge about the Hill in the constant challenge to take care of it in the face of threats of irreversible destruction from tourist developments and an uncaring local authority. There are now many more local environmental and residents' groups aware of the current value of the land and the potential loss Swansea faces if the destruction begins. This book was the first of the Cilfái trilogy, the second book covered Woodland Management and Climate Change, and the third book covered the heritage features on the hill. This book was rushed into print to address claims from the Local Authority that there was 'not much up there'. Since 2023 I've taken hundreds of people on walks to view the biodiversity, history and heritage of Cilfái, and I've packed out numerous community centres and halls talking about the history. Hundreds of people have been converted to the value of the landscape we may lose.

Contents

Introduction 9

One: Kilvey 13

Two: Coal and Kilvey 18

Three: Copper and Kilvey 37

Four: Pollution and Destruction 47

Five: Repair 62

Six: Nature and Restoration 77

Annex 1: The original White Rock Lease of 1737 93

Annex 2: Extracts from William Logan's fieldnotes 1837-40 105

Index 121

Introduction

Kilvey Hill is an iconic part of Swansea; it dominates the town centre on its eastern side. The significance of Kilvey to Swansea residents and visitors will increase as it becomes a prominent and visible green space adjoining the city's urban centre. It is a part of the identity branding of Swansea.

Kilvey's recovered woodland is a superb example of a heavily polluted landscape's post-industrial recovery; better communication of its past and future significance is needed at every opportunity. It's not perfect, and it wasn't always so green. The coniferous woodland of the 1970s can be seen to be past its best, and anyway, is now seriously out of fashion in government policy as a new crusade for 'native woodland' is championed by Welsh politicians.

Ecologically, Kilvey was once destroyed by industrial pollution. The primary source was the adjoining White Rock copper works, probably Swansea's most famous industrial history site. However, little survives above ground there to give sense to the two centuries of industrial activity. Readers may feel I have been unfairly harsh about the Lower Swansea Valley industries that played such a large part in Swansea's history. However, the cheerleaders of industrial archaeology are numerous, and many books cover the popular local industrial history extremely well, and I have acknowledged those sources freely in the text. I trained as an industrial archaeologist in the 1980s and am enthusiastic about the subject. However, the predominant theme here is Kilvey and the environment—a seriously neglected topic.

I know the area well, having grown up locally and commuted by foot, bicycle, and car past White Rock for 30 years. I fondly remember early morning encounters with rabbits and afternoon spotting of reptiles as I walked along the surviving railway tracks until the landscaping of the 1990s finally swept them away.

Although the land was never officially part of the Lower Swansea Valley Project (LSVP) of the 1960s, the green mantle of recreated woodland is undoubtedly one of the historical achievements of Swansea Council, Swansea University and the other agencies that collaborated on the restoration. The plant species planted in the 1970s were based on the knowledge gained by the remarkable botanical and chemical work of the LSV Project. The woodland is a historic part of the recovery of Swansea from the consequences of the enormously polluting industries that destroyed the Lower Swansea Valley. The predominance of Lodgepole, Corsican and Scots

Left: Central kilvey in the hot summer of 1981. At that time there were still very few trees and a large community of coarse grasses and heathers covering the thin soil. Fire risks were still low with arsonists having to work hard to get a conflagration started. (Author's collection).

pines across the woodland is because of their resilience in Kilvey's thin and toxic soils. Many of those 'pioneer' trees are now sick or old, and their work has been done. The large amount of natural native tree regeneration in the clearings and gaps on Kilvey demonstrates that the woodland's healing and recovery process is underway. The natural conversion or transition from coniferous forest to broad-leaf has already begun, and those natural regeneration processes should be supported wherever possible. The increasing momentum of natural regeneration should make any formal planting schemes an unnecessary expense.

Some original pines planted in the 1970s have grown to impressive sizes and are still safe and healthy. They should be considered historical artefacts and assets because they were part of that original pioneering effort to tackle pollution. This kind of 'post-industrial woodland' is a significant part of the landscape recovery and rejuvenation of the industrial regions of Wales. Welsh Post-industrial Woodland takes a place alongside Ancient Woodland, Ancient Semi-natural Woodland and Plantations on Ancient Woodland as essential assets of landscape recovery and adaptation to new climatic conditions.

As the woodland has grown, so have its use and connections with the communities surrounding Kilvey. A thriving volunteer organisation has provided thousands of hours of voluntary time to care for and protect the land and is one of many organisations and groups that use the woodlands for well-being and recreational activities.

This study isn't a conventional history of Kilvey. Much of the area's rich history is already documented in the various history books referenced throughout the text. This is particularly true of our industrial history, extensively researched, with some researchers even feeling there is little else to investigate. But, perceptions of what is good history change with the generations. Swansea is incredibly rich in historical enquiry with a substantial body of knowledge in journals and books, both large and small. Instead, as a geographer, I've taken a view through my lens of spatial and environmental training. These chapters cover what I've been teaching, researching and actively involved in over the past years. If I'd looked at my notes in the 1990s, I can see that I'd have most likely created a more conventional history of the land with a chronological approach underpinning the theme. More recently, I know from my students, and even my time as a government analyst, that tastes and interests have changed. The climate emergency is encouraging students and activists to ask different questions of policy and politics. Equally, the dearth of local historical teaching in school curricula has led to intense curiosity from many who want to understand and contextualise the place of their communities in the general story of Swansea's history and environment. On the plus side the demand for good quality local history information remains strong. I am always a strong believer in the adage that in history, there is less of a distinction between 'amateur' and 'professional', but rather more important is the distinction between 'trained' and untrained'. Any history or geography courses I have led over the past two decades invariably start with an introduction to Swansea's basic local history to try and begin to explain 'why we are where we are.'

My family lived and struggled with working in Hafod and Landore for three generations, so

I am a son of the industrial areas of Landore and Port Tennant. In common with others, I find associating what I know of my family's history with the burgeoning 'nostalgia' industry of Copperopolis somewhat hard to take, particularly as the layers of gloss on past events increase in response to new tourism requirements. I look at some of the new tourism developments at Landore and Hafod with a quizzical eye as I see unhelpful simplification and characterisation of past complexities, particularly in terms of pollution and environment.

As a discipline, geography has changed considerably over the past few decades. The rise in interest in environmentalism is now mainstream, putting pressure on the subject to reinvent itself to fit the demand. In my specialist field of historical geography, the questions have changed as we seek to be more helpful and 'applied' and enter debates about conservation, climate change and sustainability. Kilvey is an incredible laboratory for all of this, with a remarkable past and a highly contentious future, with local communities wanting a say in influencing the future of their hill more now than at any other time. A common question I have to field from students of all ages is why Kilvey is the way it is. Politicians and planners frequently want to simplify answers to give them the freedom to do what they want. They often display a limited understanding of the past as issues are politically simplified or are buried under an overburden of opaque planning jargon. However, local communities also want to influence that future. For both sides, understanding must start with being aware of the environment, the history, and the reality that there is a contrast between the complicated past and the complex future. For example, many people want to see more trees on Kilvey, and government policy is for trees to be native species. However, Kilvey has many coniferous trees because of the polluted past of the land, where little else would grow. A set of wide-ranging policy statements on what is needed on Kilvey will not do away with the realities of facts on the ground and recognition of the past events that shape what is there now and what could be there in the future. Explanation sometimes has to look at past events, not least to understand the processes of continuity and change in a world where timescales are constantly being shortened, and misinformation or generalisation is a common currency.

So, my geography of Kilvey is a mix of history, environment, people and place. Academically, this approach has sometimes been described as 'environmental historical geography', a long label coined by geographers who need us to have labels (Williams 1994). Although the content here is hopefully more straightforward, as it's the stories and facts I have needed to answer the student's question, 'Why is Kilvey like it is?'

This little book is a tour through a unique landscape with some remarkable history, some particular significance, and hopefully a sense of optimism and ambition that people recognise and understand the uniqueness that is our Kilvey.

I have reverted to the Welsh language name of 'Cilfái' for most of my work.

Nigel Robins, Swansea 2025

Map of Swansea area showing placenames including:

- Felin-fforest
- Hen-clau-du
- Wych Tree Bridge
- Ffynudrod
- Llansamlet
- Pen-lan
- Clas
- Morriston
- Cwm
- Ty-gwyn
- Llwy...
- Fforest-isaf
- Dyffryn-aur
- Pen-y-waun
- Trwyddfn
- Brass wire Works
- Llwyn crwn
- Tal-chopa goch
- Blaen-ay...
- Cwm-gelli
- Cnap-llwyd
- Pentre Engine vch
- Pentre Engine fawr
- Trallwyn
- Tal-chopa Eian
- Copper Works
- Colliery
- Tir bach
- Pentre cawr
- Winch wen
- Lan
- Pentre-dwr
- Llys-newydd
- Canal Rail Road
- Gwern Mostyr
- Lan goch
- Pant-y-moch
- Tir mawdy
- Lan-y-wern-uchaf
- Crymlyn Brook
- Lanstr
- Pwll-yn-aur
- Upper Bank Copper Works
- Gwern Mostyr Farm
- Pentre-grasea
- Cefn hanod
- Lan-y-wern-isaf
- Pant-y-trfeiliad
- Hafod Copper Works
- Gwern-tu Cnap goch
- Bon-y-maen
- Tir draw
- Coal-pit
- Tir dinam
- Middle Bank Copper Works
- White Rock Copper Works
- Llwyn-hatwner
- Bixley hill
- Ty coch
- Cwm dial
- Pwll-cynon
- Pentre-graig
- Fox hole
- Pen y graig uchaf
- Cwm bach
- Tir isaf
- Kilvey Hill
- Pen y graig isaf
- Tir John North
- Pentre-grance
- Caer brwyn
- Gelli-graeog
- Tir-owl
- Tan-y-grain fawr
- Canal
- Tir lan...
- Tan-y-graig
- SWANSEA
- Lodge
- CRYMLYN Race Course
- Harbour Ford
- Dr. Burrow's Lodge
- East Pier
- Port Tennant

12

One: Kilvey

Kilvey Hill is big. If you grew up on Mount Pleasant and Townhill as I did, you grow up looking at it. It is at the end of the street in Norfolk Street; in the 1960s, it was reflected in the windows of Andrew's Bakery.[1] Then, when we were children, they built a TV antenna on the hill topped with a bright red light. So even in the dark, we knew it was there.

Kilvey is a big lump of hard sandstone, part of a series of hills going westwards to the Gower peninsula along the coast. The rock is blue-grey when newly exposed but quickly weathers to a rusty brown colour when the weather gets at it. That rusty brown is frequently the colour of Swansea, at least for most buildings built up until the 1800s when brick and cement became popular. The sandstones, called Pennant by geologists, give up coal in huge quantities, of which more later (Owen and Rhodes 1969: 32–37).

Kilvey, or in the early Welsh Cilfái, was an eastern appendage of the Lordship of Gower from the 1100s.[2] The community on Kilvey has a distinguished history, best enjoyed in tracing its stories through the monolithic Glamorgan County History (Pugh 1971). Kilvey has dominated the geography of Swansea since the town's foundation because it constricts the Tawe valley as it enters Swansea Bay, forcing roads and transport links around it to the north via Llansamlet or to the south around the flat lands of Fabian's Bay. Several steep roads traverse the hill, ancient tracks that suggest travel by foot or pony rather than wheeled traffic. Morris Lane, the most important track, is still a regular route for walkers today.

It isn't easy to imagine what medieval Kilvey was like. The land was suitable for animals and crops, and there was plenty of water in clean streams that ran down to Foxhole. The medieval corn mill at the Knap Goch site (later White Rock) suggests a busy agricultural life, and of course, there was coal.

Swansea's industrial revolution started earlier than the rest of Wales because it was powered by easy access to coal. It is hard to understand the importance of coal today, particularly as it is increasingly vilified as the initial cause of climate change (Malm 2016: 13, 220, 225). Coal and Swansea were intimately linked; earlier

Left: Kilvey (Cilfay) as shown on the Ordnance Survey Sheet XXXVII in about 1830. This map is a rare early printing of the map that would be universally known as the 'One-Inch Map' because of its original 1 inch to 1 Mile scale. This map shows detail originally surveyed in about 1813 so we get a glimpse of an earlier Kilvey than the edition date would suggest. Relief is shown by hachures instead of contours This is the earliest reliable map of Kilvey placenames. The surveyors tried to recognise Welsh names and pronunciations, although later versions of the map would be far more anglicised. The Welsh 'Cilfái' becomes 'Cilfay' and ultimately 'Kilvey. (Author's collection).

generations would have talked of coal with the names of Swansea's villages and communities. The Penlan vein, the Foxhole vein, the Dyfatty vein, and the Tormynydd (Danygraig) vein were all important. Kilvey had the Hughes vein, Captain's vein, the Rotten and the Warky (Strahan 1907: 44–45). These veins of coal were worked from early in Swansea's history, of which more later. The signs of coal working mark the hillsides from the western slopes of the Afon Llan near Penllergaer across Townhill to the eastern slopes of Kilvey. Small rocky outcrops and small quarries mark old coal workings or explorations as prospectors looked to find easy access to coal.

Much of Kilvey's modern history is closely associated with the adjoining White Rock and Middle Bank copper works, which were of global importance at their peak production in the nineteenth century (Evans and Miskell 2020). The industrial archaeology of the White Rock copper works is well-known and thoroughly documented. However, it can be challenging to appreciate the technical nuances of the arcane smelting practices of early metal industries, although they do enthuse many (Hughes 2008; South West Wales Industrial Archaeology Society 1975). In addition, some lament the 1960s destruction that cleared many of the surviving ruins. Still, sometimes there is an insufficient acknowledgement of the priorities of the Lower Swansea Valley Project, which concentrated on residential and industrial reuse of the destroyed land and gave less regard to the preservation of ruins, no matter how valuable they seemed to historians and enthusiasts (Hilton 1967: 1–14).

Looking at the White Rock site today, somewhat mangled between road junctions and a large roundabout, it is impossible to envisage the dense clusters of industrial buildings crammed into that bend of the River Tawe and how it all related to the slopes of Kilvey (Hughes 2008: 329). White Rock's proximity to Kilvey meant that it was profoundly affected by every activity in the works, whether it be water extraction, water pollution, coal mining, smoke pollution, or waste tipping. Ironically, apart from the two remaining Hafod chimneys on the opposite riverbank, the most visible legacies of Swansea's copper industry nowadays are Kilvey's woodlands, the landscape scars amidst the trees beneath them, and the Nant Llwynheiernin that runs clean and clear once again.

Notes

1. The Norfolk Street Bakery of the Andrew family was one of the vital food shops in the village of Mount Pleasant for over a hundred years (Andrew 2019).

2. The more elegant Welsh spelling of 'Cilfai' has sadly been anglicised since the 1860s into the modern 'Kilvey'. The name was originally interpreted as 'Cilfay' on the earliest Ordnance Survey maps of the 1830s, however, the trend for later OS maps to reflect anglicisations of local place names has irretrievably damaged much of our understanding of past geograpphy. The finest analysis of local early geography that I have come across is that by the pioneering historian Clarence A. Seyler (Seyler 1924).

REFERENCES

Andrew, John, 'Our Daily Bread: A Century of Andrew Family Baking in Mount Pleasant', Swansea History Journal/Minerva 27, 2019-20, 30-42.

Evans, Chris, and Louise Miskell. 2020. Swansea Copper; A Global History (Baltimore: Johns Hopkins University Press)

Hilton, K. J. (ed.). 1967. The Lower Swansea Valley Project (London: Longmans)

Hughes, Stephen. 2008. Copperopolis: Landscapes of the Early Modern Industrial Period in Swansea (Aberystwyth: Royal Commission on the Ancient and Historical Monuments of Wales)

Malm, Andreas. 2016. Fossil Capital (London: Verso)

Owen, T.R., and F.H.T. Rhodes. 1969. Geology around the University Towns: Swansea, South Wales, Geologists' Association Guides, 17 (Colchester: Benham)

Pugh, T. B. (ed.). 1971. Glamorgan County History Vol. 3: The Middle Ages (Cardiff: Glamorgan County History Trust), III

Seyler, Clarence A. 1924. 'The Early Charters of Swansea and Gower Parts I and II', Archaeologia Cambrensis, Journal of the Cambrian Archaeological Association, 7, Vol. 4, (1924): 59–79, 299–326

South West Wales Industrial Archaeology Society. 1975. Industrial Archaeology Trail of the Lower Swansea Valley (Swansea: Swansea City Council)

Strahan, Aubrey. 1907. The Geology of the South Wales Coal-Field. Part VIII, The Country around Swansea: Being an Account of the Region Comprised in Sheet 247 of the Map, Memoirs of the Geological Survey. England and Wales, 247 (London: Printed for His Majesty's Stationery Office by Wyman and Sons)

Two: Coal and Kilvey

Until World War One, Kilvey and coal formed an intimate partnership bonded by coal mining from Roman times until the last coal mines on the hill's eastern side were closed in the 1920s. The signs are everywhere, from recognisable waste tips with exotic shapes best picked up by Lidar surveys to orange staining of outpouring chalybeate springs as groundwater flows through long closed tunnels (Strahan 1907: 80–81). Early coal tips from before official records began are often hard to find; however, one of Glamorgan's most competent local geologists, the common badger, often finds them as they are easier to dig into, leaving it far easier for me to find tips from the spoil around a badger sett. There's scant white colouring on a Glamorgan badger living on the coal measures.

Finding coal on the surface can be hard seventy years after Kilvey's coal mining ended. The near-ubiquitous material that dominated human activity on Kilvey for a thousand years is almost invisible. Areas where coal outcropped on the hill, which were of immense importance in past centuries, have now disappeared beneath landscaping, regrowth, and deliberate masking of the rocks. Nevertheless, as an amateur geologist, finding a coal specimen on the hill is a significant event equivalent to any fossil finding on the limestone coast of Gower.

Coal is not a mineral; it's a rock, something that has confused many over the years. Coal is a rock of vegetable origin with organic matter transformed by biochemical and physical action. Although from Roman times, people knew what coal could do for them in fires and smelting, defining and understanding how and what coal has taken more than a century of biological and chemical research from the early microscope examinations of Henry Witham in the 1830s (Witham 1831) to the milestone work of palaeobotanist Marie Stopes in the 1900s (Stopes and Wheeler 1918), and Swansea's Public Analyst Dr Clarence Seyler (North 1926: 4–6).

A lack of clarity on coal's scientific origins and nature did not affect how it was exploited. Coal was a Swansea staple for export and local use from at least the 1300s and probably much earlier, explicitly referring to the right to dig coal in the 1306 Charter granted by William de Breos and the burgesses of Swansea (Alban 1984: 12–13). From the 1650s, coal was an industrial product with massive investment in prospecting, technology, and markets, with significant impacts on the geography of Swansea (Hughes 2008:

Left: Small lumps of Kilvey coal from the site of the Hughes vein on the hillside above White Rock. Despite being ubiquitous a century ago, finding coal on the surface is rare in the coniferous woodlands. Kilvey coal is a hard coal, bituminous in character, that would give off clouds of thick smoke when burnt with Cornish copper ore (Author's collection).

2–16). Swansea's long-held experience with the winning of coal meant that local knowledge of coal, its qualities and availabilities were immense. Traditional local knowledge was hugely important to understanding where coal was and how to get it. The importance of oral tradition, observation and experience, and pooling collective knowledge was vital in an era before the laws and theories of geological succession and strata started to emerge with the work of Abraham Gottlob Werner in the mining areas of Germany and the pioneering work of William Smith in the early 1800s (Macfarlane 2020: 35–39).

We get insight into early Kilvey coal mining from an excellent source from 1400. In a short entry in a bundle of accounts relating to income for the Lordship of Gower, we have an annual report on the income and activities for a single Kilvey coal mine (Nef 1932: 422–23). Unfortunately, we only get sight of one year, and the location is only given as 'Kilvey', although that is enough to suggest the site was close to White Rock or Foxhole as it had to be near the river near the veins of coal known as the Hughes or Foxhole veins. It was a large enterprise, employing three miners and thirty porters who carried coal from the mine to the riverside in wheelbarrows for export to Cornwall, Devon, Ireland, and France. The French connection was long established by the 1500s (Alban 1984: 17). The mine produced approximately four thousand tonnes of coal in 1400, although it's difficult to say if this was a good output for the time. The operation also needed considerable investment in candles, repair and upkeep of tools, wheelbarrows, and underground tubs on sledges to move coal away from the workfaces. Water was a problem for them, and drainage and guttering were needed to deal with underground water flows. This was a drift mine, where small tunnels were cut into the sides of the hill following the veins of coal—confirmation of medieval technical expertise, geological knowledge, and economic planning. The geology of coal veins on Kilvey was particularly suited to this kind of drift mining; the hill's western side shows many rock exposures and trenches indicating extensive drift mining from the 1300s up to the 1750s.

Estimating how much coal Kilvey produced is impossible as too much went undocumented or unknown from many small mines and quarries. However, local use inevitably increased because the population increased, but most of the Kilvey coal could have been for the export market (John 1980: 25). What is more certain is that things started to change in the 1750s. Coal becomes a more valuable commodity, with coal-bearing land seeing an increase in value, particularly around northern Kilvey and access to the White Rock and Foxhole coal shipping quays and the White Rock copper works (John 1980: 25). The increase in coal use is linked to what we call 'industrialisation' today. However, the term was not used in the eighteenth century (Williams 1974: 360–61). For Swansea, the year

Right: The remains of a coal mine on the hillside above White Rock in 2022. The Hughes vein was exposed along the northern side of the course of the Nant Llwynheiernin and was heavily worked from Medieval times up until the 1750s when workings were abandoned for deeper, more productive pits in the Lower Swansea Valley. This exposure was one of several that William Logan examined closely as he taught himself the geology of Swansea's coal fields in the 1830s (Author's collection).

1760 is usefully regarded as a point where there was 'a quickening of the pulse' of local industry (Williams 1974: 361). The discussion over the date of the step change in coal use has become of greater interest recently as academics look to understand the origins of global warming, climate change, and the role of fossil fuels, particularly coal. As the 'first industrial nation', Britain's industrial revolution has attracted a 'minor flurry' of interest (Malm 2016: 13–17). Inevitably, coal and the steam engine become the focus of interest. As both have long histories in Swansea, it is interesting to reflect on a possibly more prominent role of Swansea's history in the climate crisis. Andreas Malm is not demonising Swansea (particularly) in the origins of climate change. Instead, he is seeking learn-able lessons in how the fossil fuel economy came about and reflecting on the 'unique archive' of lessons in economics and energy development that reside in our industrial history. From an environmental point of view, if coal (carbon) is a villain, then Swansea has a hefty track record worthy of examination, both in terms of local pollution and, if you follow Malm's views, in terms of the planet's current climate emergency (Malm 2016: 4–8). We see a deliberate switch locally from elaborate sustainable water power to steam power between the 1780s and the 1820s.

On Kilvey, we see an elaborate water power engineering scheme developed mainly between 1700 and 1800 with great expense and expertise and part of a more comprehensive sustainable water power network from Birchgrove in the north of Kilvey to Tir-isaf in the east, providing water wheel power over thirty collieries and industries (Hughes 2008: 107). Ultimately, up to a hundred large water wheels provided renewable energy (Hughes 2008: 124). Interestingly, by the 1830s, the proprietors of White Rock copperworks had started dumping copper smelting waste on their part of the network, as if the water network didn't matter anymore. This rapid abandonment of water power in preference for steam occurred across England and Wales throughout the 1830s, particularly in the north of England (Nuvolari and others 2011: 295–96). The abandonment of waterpower had a profound impact on Kilvey as the careful curation of water resources was abandoned, and massive waste tipping added to the ecological damage already created by copper smoke.

We have seen that Kilvey's coal history is long, albeit undocumented, with much of the knowledge passed down in oral tradition, practical experience, and traditional skills. The arrival of the White Rock copper works and the later industries meant a new 'industrial' approach to coal knowledge was needed. Coal exploitation for industry required considerable investment,

Right: My sketch map of the main features of the coal veins as explored by William Logan in the 1830s. At that time, basic knowledge of the coal veins was good, and local colliers knew Kilvey had three good coal veins (Hughes, Warky and Captain's). The Foxhole vein was harder to trace and although Foxhole coal was of fair quality, the vein was not strong and tended to peter out eastwards. Efforts were made to find a better access to Foxhole coal from trial pits at the top of the hill but nothing useful was found. The yellow colouring either side of the river is land known to be coal bearing, but the river floor coloured blue on the map had a thick layer of silt and glacial debris which meant that coal pits had to be sunk deeper and needed pumping with steam engines. This map shows what Logan thought was happening from the evidence at the time. We now know that the complex coal seam faulting across the valley area makes the actual picture more complicated. (Author's collection).

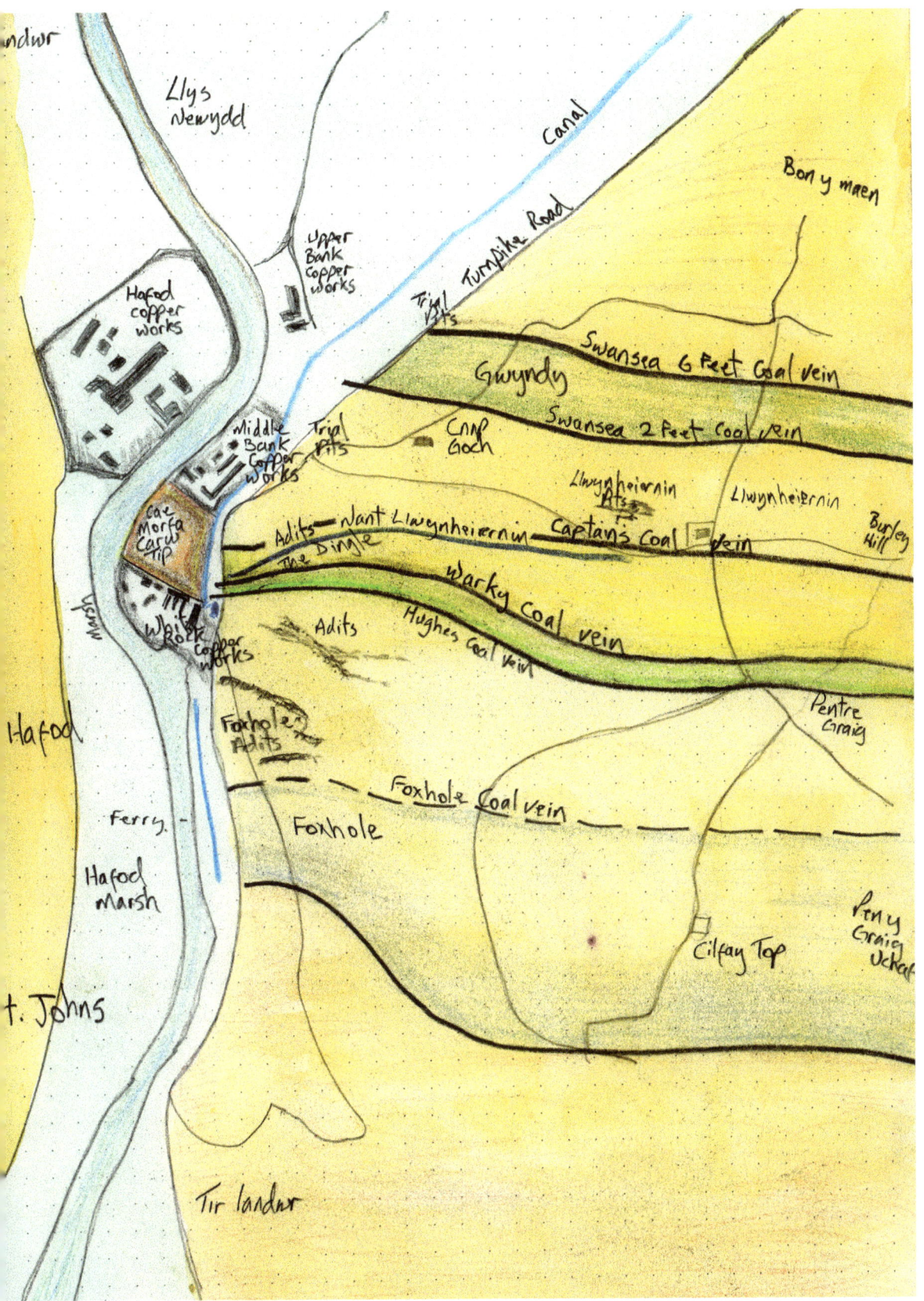

and investment required knowledge of finance, extraction techniques, and commercial certainty. One historian described the early nineteenth century as the 'age of the polymath virtuoso engineer' (Trinder 1982: 128–35), and they needed data as much as money to create wealth. Therefore, economic geology in the shape of coal information was in demand.

Understanding geology has been a neglected aspect of Swansea's history. Aside from a few general maps of 'coal seams' or references to something like the 'Swansea five-foot seam', little recognition has been given to one of the crucial building blocks of Swansea's industrial history. The winning of geological knowledge is fascinating, mainly as Kilvey was central to so much of it.

For eighteenth-century industry to grow, the oral tradition of coal mining on Kilvey and the surrounding lands of Swansea needed to be understood and translated into scientific proof of the existence and proximity of underground coal resources. The explosion of the coal trade in the 1750s brought the expenses of new steam technologies for pumping and winding and new productivity targets to feed the valley's industries whilst also increasing exports (Williams 1980: 159–63). The ability to mine deeper and deeper relied on the certainty that mines were being dug in the right areas. The drift mines of Kilvey or the Llan Valley at the Penllergare Estate followed easily identifiable coal veins on the surface. Still, they would never provide the quantities of coal that copper smelting or other industries required. The oral traditions needed to be interpreted, and a new scientific understanding of where coal lay needed to be established.

A thousand years of coal mining experience had long established some truths. Coal varied in quality depending on where across the area it was found. In eighteenth-century terms, there was stone coal, culm, bituminous, 'cokeing', and some harder 'steam coals' deemed very suitable for steam engines (Conybeare and Phillips 1822: 426–28). Doubtless, local colliers had several other names for the coal qualities they found and their reliability. It was also widely known that coal veins were interspersed with sandstones, shales, and siltstones. The famous 'Pennant Sandstone' of Swansea punctuated the coal veins and was a ubiquitous building material for houses, canals and railways before cheap bricks became available. The high iron content of the sandstone means that the blue-grey colour of the rock quickly weathers to the warm browns that make up a large part of the 'colour' of Swansea. Another essential feature of Swansea coal was faulting. The sandstones and veins of coal across Glamorgan were cracked and broken and had many faults meaning that a vein of coal could disappear from view in a tunnel and continue many metres above or below the coal face. Faulting could have a fatal impact on the viability of any coal mine. Knowing where the coal vein would eventually reappear in a tunnel

Right: An extract from Swansea's first geology map. This sheet was seen as a triumph of early geological exploration and largely created by the intense efforts of William Logan in 1840-42. The base map is the original Sheet XXXVII, which provided an accurate base on which to plot the main coal veins that crossed Kilvey and the valley. Logan's pioneering work in interpretation was based on his self-taught knowledge and considerable effort with a theodolite and a barometer to measure heights. The concentration of coal seams on Kilvey above White Rock had been of economic significance since the 1400s (Author's collection).

23

or shaft was a significant skill of the best local colliers. Beds of Welsh coal were not conveniently horizontal as they are in other parts of the world. They dip and slant in various directions so that coal seams exposed on the surface in Foxhole on Kilvey could be hundreds of metres underground in Morriston or beyond. Understanding the 'dip' of a coal vein and mapping its passage through the ground became a crucial skill of the new generation of eighteenth-century surveyors. The dip of the coal veins was often why Glamorgan hillsides and steep valleys allowed easier access to coal. However, the search for more productivity pushed industrialists underground as deep as technology would allow.

The power of traditional knowledge and oral tradition meant that these early geologist colliers are unknown to us as they were workers and not writers. As historians, we must always be aware that the surviving records of any historical period rarely supply an accurate view of the past. The history of early geology is particularly fraught with this problem. Nevertheless, we are lucky to have an early article from Edward Martin, one engineer who must have built up much knowledge and experience from older generations of colliers.[1] Martin is the first surveyor who wrote up his knowledge of the Glamorgan coal measures in a surviving record (Martin and Greville 1997). Martin's summary of expertise has been described as 'groundbreaking', and its impact lasted for a generation or more (Torrens 1999: 100–101). Crucially, Martin summarised and filtered traditional local knowledge into a more scientific form allowing an early understanding of the relationship (stratigraphy) of the coal veins he knew of in Glamorgan and communicating these wider across Britain (Williams 1980: 170). As early as 1806, he gave shape and detail to the South Wales coalfield. He undoubtedly influenced another incredible pioneer of geology William Smith, who produced one of the earliest geological maps, again influencing generations of subsequent geologists (Macfarlane 2020: 201–21). Martin would have known Kilvey very well, and he would also have appreciated that the dipping veins of coal that were seen on the hill would have been accessible further north on the Duke of Beaufort's land to the west of the Tawe and the Mansel Estate in Llansamlet.

Martin's influence on local geology was significant enough to last long after his death. For example, one of the essential British geological textbooks of the 1820s relies entirely on Martin's 1806 assessments, collated from numerous coal miners in the late 1700s (Conybeare and Phillips 1822: 426–28).

Coal mining on Kilvey continued to develop into the late eighteenth century, although older, less productive sites were quickly abandoned. The shallow pits of Llwynheiernin on the hill above White Rock appear to have been abandoned by the 1830s, leaving a few weathered coal tips as the only modern trace. The western slopes of Foxhole and the Nant Llwynheiernin still have traces of the open workings or adits used to 'chase' the coal seams up over the hill.

The continuing expansion of industry in the Lower Swansea Valley and the development of Upper Forest Works in the 1820s would bring William Edward Logan (1798-1875) to Swansea. Logan, born in Canada, was one of the most significant geologists of the nineteenth century and is now regarded as one of Canada's most influential scientists. Initially trained in chemistry and mathematics, Logan nurtured his

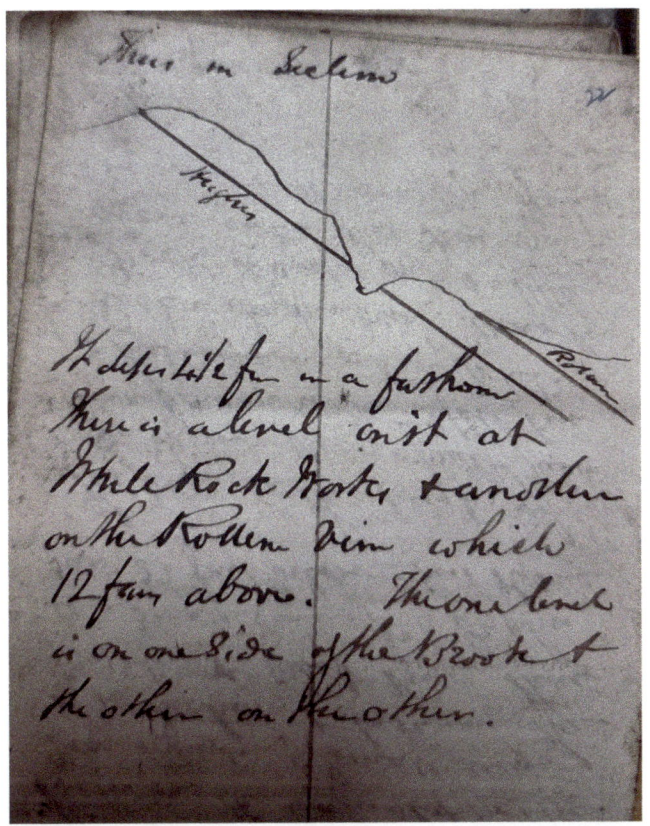

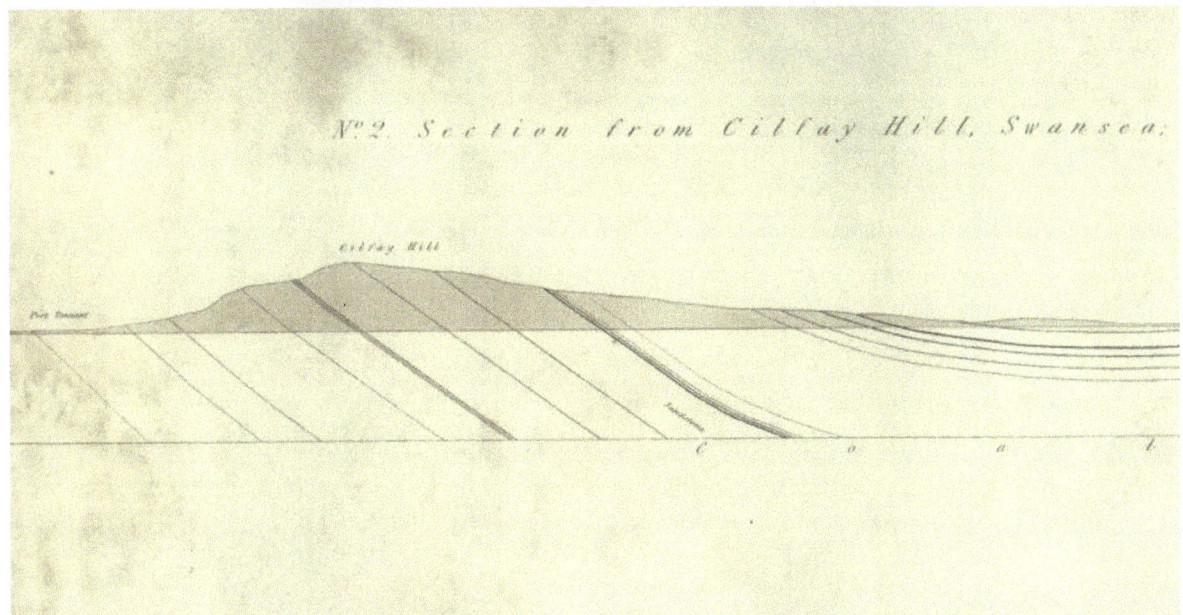

Top: An extract from Logan's notebook recording his examination of the hillside above White Rock in 1837. By this time Logan was working hard to understand the dip of the rock beds which meant that coal exposed on Kilvey was far underground at Morriston. (British Geological Survey Archive).
Bottom: Logan's original section through Kilvey from 1842. Although this looks simple, it was the first time the dip of the coal had been graphically shown and was based on Logan's intense work with theodolite and barometer and a lot of site inspections. (Author's collection).

amateur geological knowledge on the hillsides of Swansea, making massive advances in our understanding of the Glamorgan coalfields before returning to Canada to lead the Canadian Geological Survey. His work in Swansea was so influential that Logan is still acknowledged on the current geological map of Swansea ('England and Wales Sheet 257: Swansea Bedrock' 2011). Logan's early studies began in 1831 or 1832 on Kilvey Hill.

Before discussing the important work of William Logan and Kilvey, it is helpful to know a little about maps. Nowadays, we have so much access to maps via our smart phones and Google that it is impossible to imagine life without them. But creating valuable maps of Britain in the early 1800s unlocked so much of the country's development potential. The British government mapping organisation, the Ordnance Survey (OS), began initially as a war department in the 1790s. Still, the OS was quickly turned to mapping for civilian and economic use (Seymour 1980: 2–11). Although several general maps of Swansea and Gower were available in the 1700s, none were reliable at a consistent scale. The lack of an accurate base map to plan anything caused severe problems in designing canals, roads, factories, and water power. Swansea was eventually surveyed in 1825, and a map was prepared at a scale of one inch to the mile (1:63,360). So, for the most part, the first reliable map we have of Kilvey, Swansea and the Valley is a snapshot of the landscape in the early 1820s. The map, the famous Glamorganshire Sheet 37 (Glam. XXXVII), is a mainstay of historical geographers studying early Swansea. In the 1830s, the important feature of Sheet 37 was the opportunity for anyone to see the area's geography laid out on one sheet at a scale big enough to incorporate detail but small enough to cover the area. As a result, people could 'see' Swansea in front of them for the first time. The earliest understanding we have of Kilvey's landscape comes from this period.

Sheet 37 was important for Logan who arrived in Swansea as a manager at the Upper Forest Copperworks (Bayliffe and Harding 1996: 20–22). Relatively new to the area (he already had some family connections here), and his managerial work aside, William Logan was a keen amateur geologist. His work at Forest brought him into immediate contact with Swansea's prime geological asset, local coal. Logan had several other key assets to understand this new world, the Edward Martin description of the coal strata mentioned earlier, a useful geological textbook, the Phillips and Conybeare 'Outlines of Geology' (Torrens 1999: 100–101; Martin and Greville 1997; Conybeare and Phillips 1822), and most notably the latest edition of Henry De La Beche's Geological Manual with knowledge of the Carboniferous

Right: The geology map from the 1860s. The years 1840-1890 saw lots of revision work on all the maps as engineers and geologists tried to understand the coal seams of the South Wales Coalfield. Logan's early work was incredibly accurate, but later geologists were able to understand more about the complex faults that broke the coal veins up and made working coal in Swansea very difficult. You can see the faults added as white lines on this hand coloured map. This edition shows the new South Dock and coal veins across Townhill and north to Landore. It also shows the railways coming into Swansea via Brunel's Landore Viaduct. The grey areas on the map are coal bearing rocks whilst the lighter colour shows silt, sand and glacial material that covers the valley bottom and the coastal plain. (Author's collection).

27

rocks holding the Coal Measures (De La Beche 1833: 367–413). Logan was astute enough to look at his books and maps, look at the imperative to keep the Forest works running with up to sixty thousand tons of coal annually, and conclude he needed to understand far more about coal and mining than the geology books could tell him. He realised he needed to do that himself (Torrens 1999: 103–4). He appears to have started exploring the geology of coal within months of arriving in Swansea. Coal was a massive issue for the industrialists, and he immediately entered a world where access and costs of coal resources dominated conversations between local entrepreneurs (Bayliffe and Harding 1996: 15–29). Logan understood that whilst coal exposures were being busily worked across Swansea, most of the knowledge was based on customary and oral tradition or limited 'sections' which recorded rock types and veins encountered in sinking mines and tunnels, but not in a standardised style usable across multiple locations. Searching for better or cheaper coal supplies for the Forest works, he realised he needed a reliable map of how the coal veins spread across Kilvey, north Swansea and west to Morriston and Penllergaer. Logan hit on the critical problem early on; he needed to measure coal vein thickness, the dip of the beds and, crucially, how these related to the local topography of hills and mountains. The one important tool he needed to decode all of this was a theodolite to measure angles and directions, which he had acquired within months of his arrival in Swansea (Harrington 1883: 54). From then on; he could start to create relief maps of Kilvey and surrounding hills and understand the geology of the coal. As good as the first 1-inch map of Swansea was, it depicted relief by pen strokes known as hachures because contours had not yet been used as some felt they were not suitable for representing relief (Bassett 1967: 52). Hachures were good for visual effect but insufficient for scientifically understanding coal seams. Logan had to create his own relief data by surveying from scratch.

Logan had several advantages in exploring for coal. He had access to considerable knowledge about the quality of local coal types because they were being consumed in massive quantities by the Forest works, and he worked with coal every day. He could talk to many local mining surveyors and colliers who had generations of knowledge about local coal. Understanding the coal on Kilvey came from discussions with at least ten local colliers and mine workers (Logan [n.d.-a]: 31–36).[2] The notes from those conversations became powerful additions to Logan's calculations and surveys.

The landscape in 1830s Swansea was radically different to what we experience today. In post-industrial Swansea, we are used to trees, greenery and vegetation (albeit prompted by climate change). But that phenomenon has only occurred in the last fifty years. In the 1830s, Swansea had already endured a thousand years of deforestation and coal workings, and tips were everywhere. The Lower Swansea Valley was already a focus of toxic pollution killing trees, plants and animals. Logan would have seen a different open landscape where trees and woodlands unencumbered sight lines. Granted, the lands of Gower and north of Swansea were still heavily wooded in parts, but the stark industrial

Right: One of the Kilvey coal adits above White Rock, softened by vegetation over the past forty years. This one probably chased the Hughes or Foxhole coal veins up the hill in the late 1700s. (Author's collection).

landscape of Kilvey and the Valley made it far easier to survey the land. Standing on Mayhill, as Logan undoubtedly did, because he explored Town Hill as thoroughly as Kilvey, he would have seen the treeless bare bones of Kilvey opposite as they poked their way through the black smoke of the Valley pollution. The many rock exposures, quarries and coal pits were also a bonus for him, as he could inspect rock types and measure the direction of dip, instantly confirming his theories and calculations. For example, on Tuesday, 23 April 1839, he was walking around Morris Lane, Foxhole and around to the White Rock railway bridge (Logan [n.d.-c]: 38). On that day, he must have had his hammer and hand lens with him as he was inspecting the varying textures of the sandstones he was seeing.

Although we can't be precise with dates, we can safely assume that Logan spent much time between 1832 and 1839 on Kilvey, balancing long hours at Forest Works with long walks around Kilvey (Cilfay at that time), Llansamlet and probably further west to Town Hill and Cockett (Cockit at that time). These were the economic hotspots in which important coal veins were close to the surface or exposed. Understanding the relationship between these veins and the underlying faults in the ground enabled him to predict where coal would be. It was a constant concern. The Benson and Logan Company had a small coal pit close to Forest works where coal was barely 21m (about 12 fathoms) from the surface. Logan writes in his notebook that in the early 1800s, the owners of the nearby Calland coal pit bored through about 180m (about 100 fathoms) of hard rock and clay and found nothing. Still, a few hundred metres north of Forest, Sir John Morris's men bored 146m (about 80 fathoms) and struck the rich Church vein of coal (Logan [n.d.-b]: 21). Fortunes were won and lost.

Walking down into the valley from his workplace at Forest, the western slope of Kilvey above White Rock dominated his view. A coniferous forest today, but in Logan's time, it would have been open land stained black with copper smoke pollution and scarred along its entire length, north to south, with coal workings. Some of the workings were hundreds of years old and starting to fail as they became worked out. Many early tunnels and adits remain today, snaking their way up the hill as miners tried to follow the coal. To the young geologist, it was evident that the hill had several important veins of coal running west to east across the summit. That was an easy conclusion confirmed further by a conversation with a local collier Elias Jenkins. Jenkins was able to say that Kilvey had three good veins of workable coal; the Hughes vein, Captain's vein and the Foxhole vein (Logan [n.d.-a]: 34). Of these, the Hughes vein was the most productive and is still recognised across the region on modern maps ('England and Wales Sheet 257: Swansea Bedrock' 2011). In the early 1800s, the Captain's and Foxhole veins were also productive. The eighteenth-century efforts to chase the Foxhole and Hughes veins above Foxhole have left a snaking chasm that still exists meandering up amongst the coniferous woodland. In those days, efforts to identify the different coal veins through fossil content were still in their infancy. Despite the best efforts of local colliers, it was hard to identify the coal veins and distinguish between them.

There had been attempts to dig into the top of Kilvey to find the rich Hughes vein, but the coal found was poor. This must have been an early conversation as Logan spent much more time looking at the hill above White Rock where Nant Llwynheiernin (White Rock Brook in his notes) had rock exposures and several adits chasing coal up the hill. Logan concluded that the sides of Nant had easier access to the Hughes vein and that earlier Colliers were chasing the Foxhole as a misidentification of the vein (Logan [n.d.-a]: 35). Logan's re-interpretation of the arrangement of the seams confirmed his views (Logan [n.d.-a]: 36). The issue did not end there, Logan is back on several occasions in 1839 looking at veins and outcrops on the east and south of the hill. When he learned that colliers had dug a level into the hill near Cwmbach looking for the Hughes Vein from the east and found nothing, he knew he was looking for a significant fault in the rock of the hill. Logan examined the large stream at Burley Hill house and noted a relationship with springs at Ty Coch, Pen y Graig Uchaf, and Ty Draw further north. It was a fault line. Logan had found the Great Bon-y-maen Fault, which caused so much disruption to coal pits further north around Llansamlet Church (Logan [n.d.-c]: 37, [n.d.-a]: 31).

Logan's surviving field notes and files, preserved at the British Geological Survey archives in Nottingham, give us a helpful insight into his research. Between 1831 and at least 1837, he was full-time at the Forest copper works, but he must have spent all his spare time working on geology. He started theodolite surveys in 1832, taking height measurements checked with a barometer. He must have had assistants that helped with this, but they aren't recorded. He had two types of pocket notebook, small ones (3in by 4in) for observations and interview notes, and larger survey notebooks pre-printed with columns for the surveyor's theodolite observations. His field notes were then transcribed in ink into a large letter-book (a standard nineteenth-century office book), having details of all his work sites in the Swansea area. His notes show a gradual evolution of his geological knowledge as his skills developed. General descriptions gradually turn into more scientific detail as he examines rock structures and types with a hand lens, compass, and clinometer. Logan's careful collection of the stratigraphy of key sites recorded as coal mines and trial pits were sunk across north Kilvey and into the valley (he called them 'sections') show a methodical analysis of available information (Logan [n.d.-a]: 29). We also see the beginning of his data collection that led him to the realisation that the traditional skill of miners in understanding the nature of the fireclay that underlaid Welsh coal seams (known locally as 'undercliff') pointed to a wider phenomenon where underclays or fireclays were an important element in understanding coal seam development across the world as Logan himself was able to confirm in 1842 (Logan 1842). This discovery, first interpreted by Logan locally, became a significant contributor to the coal seam cyclothem model of the twentieth century, guiding how we understand the nature of planetary coal seams (Simpson 1969: 226–40).

As mentioned earlier, the stream called White Rock Brook (my Nant Llwynheiernin) attracted considerable attention from Logan and was a key to unlocking the coal vein pattern under Kilvey. The prominence of the Nant in the history of sustainable water power for White Rock is discussed in the other chapters. Despite the incredible upheavals and traumatic impacts of the pollution from White Rock and Middle Bank and the thousands of tons of smelting slag that covered the stream, some 200m of the original features that Logan surveyed in the stream valley have survived. Amongst the survivals is 30m of rock exposures and workings along the side of the stream that must have accessed the Hughes Vein, as Logan noted in 1837 (Logan [n.d.-a]: 36).

Logan's subsequent career as a famous geologist has been documented elsewhere, perhaps most notably in the late 1990s (Torrens 1999). However, it is hard to avoid the feeling that Logan's local work and his self-taught geological methods were overshadowed by the arrival in December 1837 of that other giant of early geology, Henry De la Beche (De La Beche 1846; North 1932, 1934). By the time De La Beche arrived in Swansea to begin mapping the local coal measures, Logan had already been mapping coal for over six years on his survey copies of Sheet 37. Logan's work was of such a high standard that, on seeing his draft maps exhibited in Liverpool in September 1837, De La Beche immediately co-opted Logan onto the Swansea mapping team as a volunteer ('Coal Basin of South Wales' 1837). Logan's work, including the Kilvey geology was published in 1843-44 on the first edition of the geological survey map for Swansea ('The South-Western Part of the Glamorgan Coalfield - Neath, Kidwelly, Swansea, Gower' 1844). Logan worked on these maps until his departure to assume the director role of Canada's Geological Survey in 1843.

The other significant part of Logan's work, which impressed De La Beche most, was his section work. All the hours of the theodolite survey and the hundreds of observations over extended lines across the Swansea and Llanelli area gave Logan a massive bank of data to develop his underground coal seams models. This work mattered economically as it allowed engineers to estimate the best locations for sinking new coal mines and understanding the rock layers before committing money and resources. Logan also identified and evaluated the impact of faulting in the rock, the one problem that massively complicated coal mining in the western part of the coalfield. His presentation to the British Association in 1837 was mainly composed of descriptions of the nature of faults around Swansea and how they impacted the coal veins. Kilvey Hill was the centrepiece of one of his sections in a cross-section through the coalfield from Port Tennant running northwards to Carreg Cennen Castle. The Kilvey section established a model for understanding the more complicated fault arrangements north of the Hill and on the western side of the River Tawe. Logan's ten years of geological work on South Wales coal, particularly Swansea, established knowledge of the landscape we still use today. Still, it is good to identify Kilvey Hill's role in Logan's career.

Notes

1. Edward Martin (1762/3-1818) was an engineer originally from Cumberland but later became the chief mining agent for the Duke of Beaufort, whose landholdings in Swansea carried extensive reserves of coal. In that role he gained extensive knowledge of the economic geology of Swansea.

2. In Logan's notes (1836-42), he refers to local interviews with Evan Benjamin, C.H. Smith, Mr Mills, Elias Jenkins, Evan Daniel, Mr David, and

historical entries from Edward Martin and C. Townsend.

REFERENCES

Alban, J.R. 1984. Swansea 1184-1984, 1st edn (Swansea: Swansea City Council)

Bassett, Douglas A. 1967. A Source-Book of Geological, Geomorphological and Soil Maps for Wales and the Welsh Borders (1800-1966) (Cardiff: National Museum of Wales)

Bayliffe, Dorothy M., and Joan N. Harding. 1996. Starling Benson of Swansea (Cowbridge: D Brown & Sons)

'Coal Basin of South Wales'. 1837. The Cambrian (Swansea), p. 3

Conybeare, William Daniel, and William Phillips. 1822. Outlines of the Geology of England and Wales: With an Introductory Compendium of the General Principles of That Science, and Comparative Views of the Structure of Foreign Countries ... (W. Phillips)

De La Beche, Henry T. 1833. A Geological Manual, 3rd edn (London: Charles Knight)

———. 1846. Memoirs of the Geological Survey of Great Britain, and of the Museum of Economic Geology London (London: Longman, Brown, Green and Longmans), I

'England and Wales Sheet 257: Swansea Bedrock'. 2011. !:1:50,000 Geology Series (Keyworth: British Geological Survey)

Harrington, Bernard J. 1883. Life of Sir William E. Logan, First Director of the Geological Survey of Canada. Chiefly Compiled From His Letters, Journals and Reports, With Steel P (London: Sampson, Low, Marston, Searle, and Rivington)

Hughes, Stephen. 2008. Copperopolis: Landscapes of the Early Modern Industrial Period in Swansea (Aberystwyth: Royal Commission on the Ancient and Historical Monuments of Wales.)

John, Arthur H. 1980. 'Introduction: Glamorgan, 1700-1750', in Industrial Glamorgan, Glamorgan County History (Cardiff: Glamorgan County History Trust), V, pp. 1–46

Logan, William E. 1842. 'XXIX.—On the Characters of the Beds of Clay Immediately below the Coal-Seams of South Wales, and on the Occurrence of Boulders of Coal in the Pennant Grit of That District.', Transactions of the Geological Society of London, 6.2 (The Geological Society of London): 491–97 <https://doi.org/10.1144/transgslb.6.2.491>

———. [n.d.-a]. "Geological Summaries 1836-1842 ' Letterbook' (Keyworth, Nottingham), British Geological Survey, GSM/GX/Lo/1

———. [n.d.-b]. 'Notebook 1' (Keyworth, Nottingham), British Geological Survey, GSM/GX/Lo/1

———. [n.d.-c]. 'Notebook 4' (Keyworth, Nottingham), British Geological Survey, GSM/GX/Lo/1

Macfarlane, Robert. 2020. STRATA: William Smith's Geological Maps, 1st edition, ed. by Oxford University Museum of Natural History (London: Thames and Hudson Ltd)

Malm, Andreas. 2016. Fossil Capital (London: Verso)

Martin, Edward, and Charles Francis Greville. 1997. 'XVII. Description of the Mineral Bason in the Counties of Monmouth, Glamorgan, Brecon,

Carmarthen, and Pembroke. By Mr. Edward Martin', Philosophical Transactions of the Royal Society of London, 96 (Royal Society): 342–47 <https://doi.org/10.1098/rstl.1806.0019>

Nef, John Ulric. 1932. The Rise of the British Coal Industry (London: G. Routledge)

North, F.J. 1926. Coal and the Coalfields in Wales (Cardiff: National Museum of Wales)

———. 1932. 'From the Geological Map to the Geological Survey', Transactions of the Cardiff Naturalists' Society, LXV: 41–115

———. 1934. 'Further Chapters in the History of Geology in South Wales', Transactions of the Cardiff Naturalists' Society, LXVII: 31–103

Nuvolari, Alessandro, Bart Verspagen, and Nick von Tunzelmann. 2011. 'The Early Diffusion of the Steam Engine in Britain, 1700–1800: A Reappraisal', Cliometrica, 5.3: 291–321 <https://doi.org/10.1007/s11698-011-0063-6>

Seymour, W. A. (ed.). 1980. A History of the Ordnance Survey (Folkestone: W M Dawson and Sons)

Simpson, Brian. 1969. Rocks and Minerals (London: Pergamon)

Stopes, M.C., and R.Y. Wheeler. 1918. Monograph on the Constitution of Coal 4 Dept. Sci. Industr. Research,' Pp. 1-58, Plates I-III. London, 1918. (London: The Society of Chemical Industry), pp. 1–58 [accessed 2 February 2023]

Strahan, Aubrey. 1907. The Geology of the South Wales Coal-Field. Part VIII, The Country around Swansea: Being an Account of the Region Comprised in Sheet 247 of the Map, Memoirs of the Geological Survey. England and Wales, 247 (London: Printed for His Majesty's Stationery Office by Wyman and Sons)

'The South-Western Part of the Glamorgan Coalfield - Neath, Kidwelly, Swansea, Gower'. 1844. Geological Survey of Great Britain (Geological Survey of Great Britain)

Torrens, H S. 1999. 'William Edmond Logan's Geological Apprenticeship in Britain 1831-18421', Geoscience Canada, 26.3: 97–110

Trinder, Barrie. 1982. The Making of the Industrial Landscape (London: Dent)

Williams, John. 1980. 'The Coal Industry 1750-1914', in Industrial Glamorgan, Glamorgan County History (Cardiff: Glamorgan County History Trust), V, pp. 155–209

Williams, Moelwyn I. 1974. 'The Economic and Social History of Glamorgan 1660-1760', in Early Modern Glamorgan from the Act of Union to the Industrial Revolution, 1st edn, ed. by Glanmor Williams (Cardiff: Glamorgan County History Trust), IV, pp. 311–73

Witham, Henry Thornton Maire. 1831. Observations on Fossil Vegetables Accompanied by Representations of Their Internal Structure as Seen through the Microscope (Edinburgh : T. Blackwood) <http://archive.org/details/observationsonfo00with> [acc. 2 Feb 2023]

Right: An extract from one of Logan's larger letter books discussing the coal measures around Llansamlet. Logan has recorded an interview with Evan Daniell of Ty Gwyn and has recorded the depths of the coal veins around Llansamlet Church. He has recorded depths in fathoms. These notes illustrate the depth of knowledge Logan needed to construct his maps and sections of the area. (British Geological Survey Archive).

Llansamlet

Information obtained from Evan Daniell of Ty gwyn formerly agent to _____ on the Llansamlet Coal district.

The perpendicular distances between the chief seams are

From Church to Middle Vein   55 fms
Middle to Great   77 –
Great to Three feet   8 –
Three feet to Two feet   18 –
Two feet to Rotten   108 –
Rotten to Hughes   12 –

The Level course of the Charles & Church pit seams is from N 45 W & S 45 E. —

Between the Engine fawr fault & the great Llansamlet fault there are two others. The one nearest the Engine Fawr throws the Coal 15 fathoms up to the east. The other throws it up to the east also, but only a few yards. —

Between the two great faults the Vein dips 1 in 10. —

At the air pit near Dyffryn aur, which is on the edge of the great fault, the Cwm level is lower than the Coal on the east side of the fault. The Level & the Coal cross one another about 50 yards farther back in the level & a proportionate number down the slope of the Vein. The level course of the Coal from the point of crossing went to about 10 or 12 yards North of the Summer House.

There is a joint to East of the Strait road to Trallwyn throwing the Coal up 5 feet.

From Engine Fawr there is a horse way down to the great Vein near the Halfway house 1300 yards. —

Between the Church & middle Vein there is a small Vein of 20 Inches & between the middle & great Vein there is a small Vein of 9 Inches. It is 22 fathoms below the middle Vein.

The Vein worked on the West of the Engine fawr fault near Bon y maen is the 2 feet Vein & that cropping out near White Rock is Hughes Vein. The lumpy Vein on the top of Cefn hill is

Three: Copper and Kilvey

A significant chapter of the historical geography of Kilvey is closely tied to the White Rock copper works. In the post-industrial era, the works have almost entirely disappeared, leaving a few vague remains and a lot of verbal gymnastics and interpretation to explain the significance of what now remains to new generations of residents and visitors. A far from easy task. The most courageous recent interpretation was from 2012 amongst a flurry of documents attempting to underpin a tranche of funding and investment rounds for development (Goskar 2012). More documents and plans are due as funding is promised to reposition Swansea's industrial heritage as a significant tourism destination. A regular promise made repeatedly from the early 1980s. The Welsh heritage documentation industry is remarkably productive, perhaps trying to make up for the fact that so much of the physical remains have been demolished or cleared because of pollution clean-ups. A CADW report of 2011 manages to devote sixty pages to Welsh industrial history without once mentioning pollution or the environment. However, thankfully, the more perceptive 2012 report mentioned above powerfully recognises the issue as part of Swansea's local heritage (Cadw Pan-Wales Heritage Interpretation Plan Wales – the First Industrial Nation 2011; Goskar 2012: 24).

The industrial development of the White Rock copper works site was best chronicled in the early 1980s by formidable historian R. O. Roberts and largely based on the surviving legal sources in local archives (Roberts 1981). It's also celebrated in the Royal Commission's massive study of Swansea's industrial landscape (Hughes 2008). It's a story of two centuries of building, rebuilding, and erratic financing reflecting the ebb and flow of capital, and foreign and domestic markets driven or disrupted by war. The inevitable technological changes in either smelting or changing the way copper was used. It's also a tale of shoehorning industrial development into a tiny corner of a riverbank constrained by a river on one side and a steep hill on the other, illustrating the worst possible consequences of industrial inertia. The problems were exacerbated as the industry expanded with new technology, production techniques, polluting by-products, and waste management,

Left: Copper slag from the early phase of White Rock smelting. For the first hundred years of operations, the furnaces produced hundreds of tons of toxic multi-coloured slag usually in small sized lumps. The early Welsh smelting process always left a certain amount of copper in the slag as is shown by the green staining top right. The production process involved throwing shovel loads of charcoal onto the molten mass to create the slag reaction and many of the slag lumps include charcoal amongst the bubbles. This is the bulk of the mound of slag that has buried Cae Morfa Carw north of White Rock and under the glade on the hillside. It is sobering to think of the thousands of wheelbarrow loads of this kind of slag that were taken up the hill by women and children (Author's collection).

eventually destroying the site's viability, but not before wreaking ecological destruction on its surroundings. Enthusiasts understandably revel in the minutiae of industrial form and function in celebrating the fragments of industrial buildings that have survived. It is a harder sell to a modern community that wants more environmentally-based heritage to enjoy,

At times the group of Swansea valley copper works produced over 70% of the world's copper, a claim quickly relinquished as the Swansea industry lost its lead in scale, technology, and markets in the 1860s (Evans and Miskell 2020: 139–57). However, despite the undoubted significance of White Rock, reliable sources for the site's early history are frustratingly scarce, with just two copy leases from c. 1736 and 1805.[1] So, although being such a significant site, the chronology from the early sources is slightly unclear. However, those records suggest that the local landowners (the Mansel Estate) were keen to encourage a coal-using enterprise on the site of Knaploth Mill, a grist mill on the side of Kilvey Hill already old by the 1730s. The eighteenth-century Mansel estate was a clever industrial operator and already had a long-established lucrative trade moving coal from mines on Kilvey down a riverside track to a small dock known locally as the White Rock Wharf for export downriver to France and beyond. The placename eventually moved two hundred metres north to become the name of the new copper works in the later 1730s.

The investment capital for such a new enterprise lay across the Bristol Channel in the affluent port of Bristol. The Mansel Estate was adept at managing its traditional coal and timber production expertise for important clients, including timber for the Royal Navy (Williams 1974: 361–62). Further afield, wealthy Bristol merchants were already familiar with the abundance of coal in the Swansea and Neath valleys and understood the economic laws that dictated the necessities of siting copper production on the Welsh coalfield (Newell 1997: 658–59). Discussions with suitable Bristol-based entrepreneurs about renting the narrow riverside site resulted in an Agreement For Sale on 24 August 1735 and a lease dated on or before 1 November 1735 with a term commencement date of 25 March 1737 for 51 years. Many other short-term leases followed in later years, but none appear to have survived. It is interesting to reflect on the sophistication of early eighteenth-century estate managers in marketing the site to various parties outside the immediate area, displaying considerable connectivity between early industrial Swansea and the wider Bristol Channel and western European economic area. We know the occupation of the land by the new leaseholders began on or around March 1737. The original analysis of the copy leases understandably concentrates on developing the copper works itself, but a re-reading for landscape history gives

Left: An enlargement of the 1744 drawing usually attributed to Thomas Lightfoot (or Lighton) . This was originally used as an illustration in the 1881 history of Swansea copper by George Grant-Francis. Normally, people look at the buildings of White Rock to the right, but as you can see here, the valley of Nant Llwynheiernin, is seen on the left along with the original corn mill and the mill pond (numbered 4 and 5). The artist has also included a large dead tree at the bottom of the valley. It is also notable that the river bank shows the encroachment of copper slag onto the marshes and meadows of Cae Morfa Carw. (Francis 1881).

some extra insights.

In the 1730s, the river bend destined for the copper works site was already busy. There was a rich salt marsh used as pasture (Morfa Carw adjoining Cae Morfa Carw), a busy tramroad down to the Mansel coal yard (the original 'White Rock'), and a large watermill with a millpond. By the 1730s, that mill was already old, described in the lease as an 'old decayed or ruinous Water Corn Grist Mill called Knaploth Mill' (a corruption of Craig Cnap Coch or similar). The lease also gave provision for the demolition of the Mill as:

'...free liberty to pull down and destroy the said Mill and to erect and build in the stead thereof any other Mill or Mills Engines or other devices whatsoever if wanted or necessary for the use of the said intended Works. '[2]

This is a nice confirmation of the adaptive reuse of the waterpower resource from Kilvey Hill, clearly intended to adapt what existed to supply waterpower to the new industrial site. Such adaptive reuse of waterpower was practically a technological necessity in early industrialisation (Malm 2016: 84–90). Generally, almost all suitable waterpower sites in Swansea had some mill activity in pre-industrial times, and industrial revolution developers were adept at recycling and adapting prior sites to new uses (Hughes 2008: 104). It shows the power of industrial inertia that a large and sophisticated specialist smelting industrial enterprise originated in a water-powered grain mill from the sixteenth century or earlier.

The Knaploth grain mill survived long enough to be included in a copy of a drawing of White Rock dated 1744 attributed to an unknown artist Thomas Lightfoot (or possibly Lighton), and shown as a substantial building in front of an impressive millpond, with a tantalising view of the stream and a valley winding up Kilvey Hill.[3] Sadly, the stream does not have a name that has survived. William Logan noted it in his coal explorations of the 1830s, referring to it as White Rock Brook or 'the Dingle' (Logan [n.d.]: 32,36). I often refer to it as Nant Llwynheiernin in recognition of the area on the Hill near Llwynheiernin Farm and Coal Pits, where it originated before the substantial alterations to the stream structure in the 1740s.

The 1730s sources confirm that the availability of waterpower and the ability to improve its quantity and reliability was a significant incentive to make the White Rock site attractive to potential tenants. The leases had substantial legal protections for the existing water networks. The Mansel Estate was keen to encourage the development by making it easy for Thomas Coster and his Bristol venture capitalist partners to exploit as much industrial waterpower as possible from Kilvey.

The earliest legal provisions to allow enhancement and development of the waterpower running into the site via Nant Llwynheiernin

Right: Industrial remains still abound in the woodlands on the hill above White Rock. The coal workings were mostly obscured in the 1970s ploughing by the Forestry Commission, but the eighteenth-century water power leats were too big to be covered up. Twentieth-century water table changes have meant that most of the industrial water features are now dry most of the year although they do retain moisture within the dry coniferous forest. Several leats have been re-purposed into cycle and walking tracks. (QGIS generated map).

Site of Knap Goch House

Reconstructed Landscape

Coal workings

The Glade

Reconstructed stream course

Original ancient stream course

Coal Adits

Waterpower Leat

Waterpower Leat

Coal Adits

Foxhole Vein coal workings

suggest forward-thinking and concern over water resources, probably because the water supply from Nant Llwynheiernin was starting to diminish as a result of extensive coal mining on the eastern side of the hill affecting the water table and reducing groundwater. However, this may reflect clear, common-sense eighteenth-century thinking on the viability of sustainable water power. The 1740s saw a lot more activity to boost the water supply of Nant Llwynheiernin by an elaborate series of water engineering developments south of the Nant to intercept water from streams running down the southwest of Kilvey into Foxhole and channel into the Nant to feed that all-important millpond. Waterpower in early industry inherited all the worries and risks of the grain and wool industries from earlier centuries. Winter freezing or summer droughts could paralyse water-powered activities and even damage the wheels and equipment. The concern that water supplies could decline (probably because of Kilvey's coal mining) increased worries further. The hammers and wheels of White Rock needed a predictable and reliable water supply to support year-round activities. The need for water was a powerful incentive to examine any industrial site's characteristics thoroughly, hence the tendency for recognised waterpower sites to become redeveloped by successive activities over time. The concerns were enough to generate a series of insurance clauses in the leases to allow the leaseholders to relocate their hammers to a more reliable waterpower site upriver from White Rock nearer Llansamlet should problems arise.[4] The efforts to engineer a reliable and sustainable water supply from Kilvey Hill were elaborate and took considerable effort, expense, and thought between 1737 and the early 1820s.

It isn't easy to estimate when the mill was finally demolished. The vigorous growth of the White Rock works in the late 1700s must have meant the building was removed and the material recycled into the numerous buildings within the works by the early 1800s at the latest. However, the millpond and associated water channels were more durable and survived as structures mapped on the 1879 edition of the 1:500 OS plan.[5] The pond probably served some useful purpose in the works as it would have been removed earlier, being in a central location on such a cramped and confined industrial site.

The original stream supplying the mill must have been quite a torrent if the present-day valley remains are an indication. The steep valley provided a significant head of water that flowed swiftly. Initially, the stream drained a large part of the north-western side of the Hill, with other drainage channels eventually built to increase water supplies as White Rock copperworks grew in its energy needs. The stream quickly became incorporated into the more comprehensive waterpower network for the area, created to service the extensive coal mining industry (Hughes 2008: 103–25). As mentioned earlier, the valley was also enlarged as it proved a

Right: An aerial view of the Lower Swansea Valley taken by German air force combat reconnaissance in February 1941 after the 'Three Nights' Blitz'. The survey aircraft covered a lot of the valley on a cold sunny morning. The Luftwaffe took some of the finest photos of the Lower Swansea Valley that have yet come to light. Because of the nature of the film and camera used, urban areas look lighter as they reflect more light, the darker areas of land to the upper right of the photo show barren land bereft of significant vegetation. I doubt there is a single tree east of the river in this photo. White Rock and the Kilvey tip are in the centre. (Author's collection).

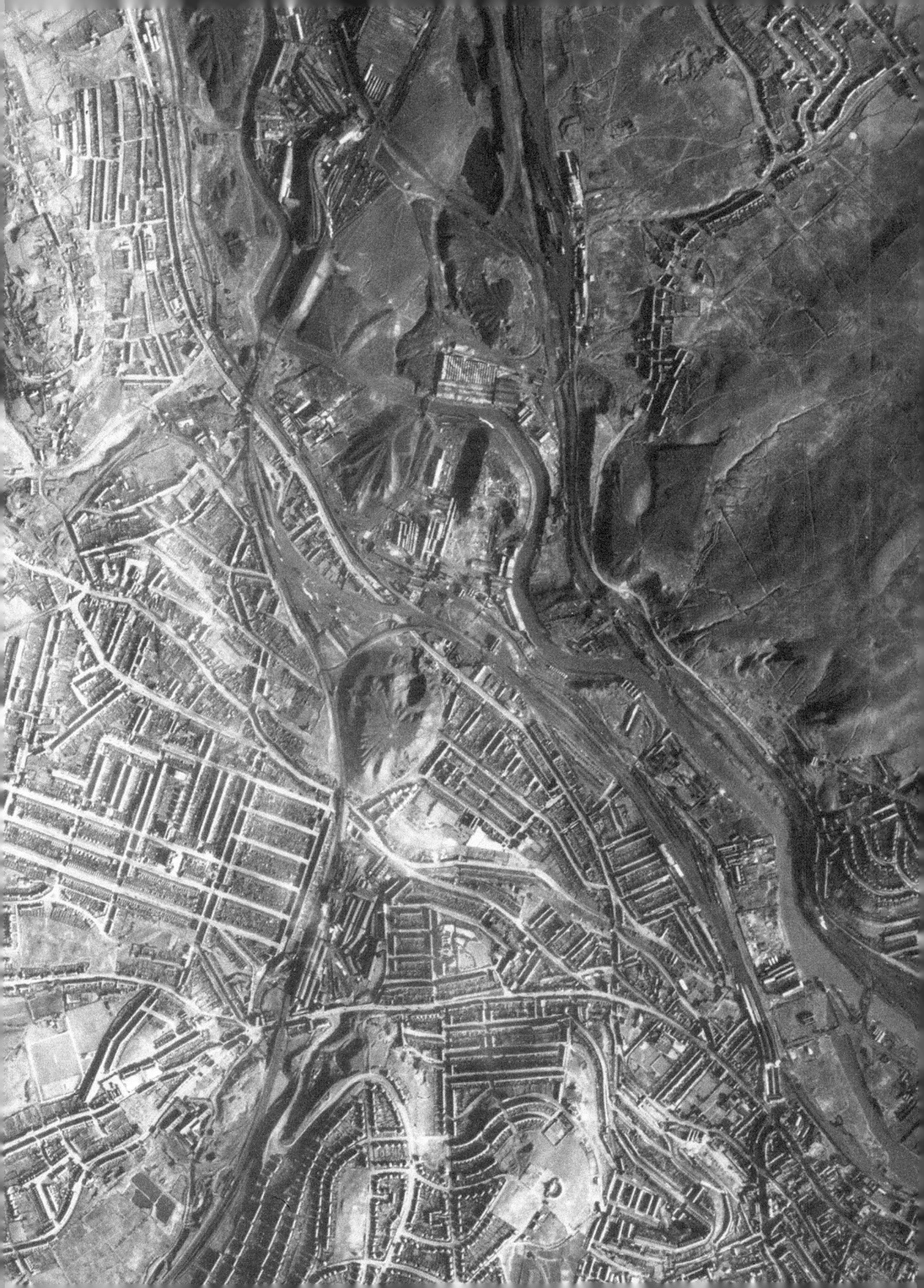

useful way to dig down to the Hughes coal vein. The valley still survives as a significant feature, although changes to Kilvey's water table mean it is barely a trickle of water today.

The critical part of the waterpower infrastructure for the early White Rock works was the channel or sluice into the Knaploth Millpond. It was probably something that, once established, would be hard to alter. This may be why the Nant Llwynheiernin inlet into the White Rock site was (and still is) a permanent fixture. The sluice survives much as it was built to carry the water under the Turnpike Road and onto the White Rock site. It currently carries the much-reduced Nant Llwynheiernin stream remnant under the Pentrechwyth road (the old Turnpike), the A4217, and into the White Rock heritage site. It then travels south in various tunnels to run over the little aqueduct at the mouth of the Smiths Canal tunnel and into the river. The survival of the sluice is remarkable, given the upheavals to the landscape in the area since the 1920s. The plan evidence shows that the stream ran in a tunnel under the Turnpike and then in an open channel into the White Rock site. The structure was mapped in 1879 on the 1:500 town plan, although its brick arch construction suggests an earlier date. The open channel survived into the 1880s on various OS plans, although it is now firmly submerged beneath the modern roads.[6]

It is incredible to think that during its lifetime (1737-1924), the site was so productive, considering it was hemmed in by the river on the west, the massive slag tip of Cae Morfa Carw to the north, various roads, canals and railways to the east, and the curve of the river to the south. Providing a suitable outlet could be found for the thousands of tonnes of industrial waste, which of course, was the slopes of Kilvey.

Notes

1. ('Abstract of Lease Dated 21 March 1736' 1736; 'Copy of Lease Dated 21 September 1805' 1805).

2. ('Abstract of Lease Dated 21 March 1736' 1736: 2).

3. (Francis 1881: 116; Roberts 1981: 138) The origins of this drawing are obscure and it's possible that the version we see today is a copy from a now-lost work. Thanks to Gerald Gabb 2022, for this extra information.

4. ('Abstract of Lease Dated 21 March 1736' 1736: 9) "...and if the water which then was or Co(uld) be brought to White Rock afsd should prove insufficient to supply a Battery Mill or Trial Hammer for proving of Copper that then it should be lawful for them to have & enjoy the mill of the sd B. Mansel called the New Mill in the parish of Lansamlet afsd the remainder of the term rent free and the same to pull down and in the stead thereof to erect & build a Battery Mill or Trial Hammer for proving of Copper".

5. (Ordnance Survey of Wales 1879)

6. (Ordnance Survey of Wales 1879)

REFERENCES

'Abstract of Lease Dated 21 March 1736'. 1736. , West Glamorgan Archives Service, DD Xhr 31

Cadw Pan-Wales Heritage Interpretation Plan Wales – the First Industrial Nation. 2011. (Perth: CADW) <https://cadw.gov.wales/sites/default/

files/2019-04/First_Industrial_Nation.pdf> [accessed 1 December 2023]

'Copy of Lease Dated 21 September 1805'. 1805. , West Glamorgan Archives Service, DD Xhr 32 [accessed 23 September 2021]

Evans, Chris, and Louise Miskell. 2020. Swansea Copper; A Global History (Baltimore: Johns Hopkins University Press)

Francis, George Grant. 1881. The Smelting of Copper in the Swansea District of South Wales from the Time of Elizabeth to the Present Day, 2nd edn (London: Henry Southeran)

Goskar, Tehmina. 2012. 'Cu @ Swansea: An Industrious Future from an Industrial Past' <https://www.academia.edu/1969069/Cu_at_Swansea_An_industrious_future_from_an_industrial_past> [accessed 27 November 2021]

Hughes, Stephen. 2008. Copperopolis: Landscapes of the Early Modern Industrial Period in Swansea (Aberystwyth: Royal Commission on the Ancient and Historical Monuments of Wales.)

Logan, William E. [n.d.]. ''Geological Summaries 1836-1842 ' Letterbook' (Keyworth, Nottingham), British Geological Survey, GSM/GX/Lo/1

Malm, Andreas. 2016. Fossil Capital (London: Verso)

Newell, Edmund. 1997. 'Atmospheric Pollution and the British Copper Industry, 1690-1920', Technology and Culture, 38.3 ([The Johns Hopkins University Press, Society for the History of Technology]): 655–89 <https://doi.org/10.2307/3106858>

Ordnance Survey of Wales. 1879. 'Glamorgan XXIV.1.14' (London: Ordnance Survey) <http://hdl.handle.net/10107/4715288>

Roberts, R.O. 1981. 'The White Rock Copper and Brass Works, near Swansea, 1736-1806', Glamorgan Historian, 12: 136–51

Williams, Moelwyn I. 1974. 'The Economic and Social History of Glamorgan 1660-1760', in Early Modern Glamorgan from the Act of Union to the Industrial Revolution, 1st edn, ed. by Glanmor Williams (Cardiff: Glamorgan County History Trust), IV, pp. 311–73

Four: Pollution and Destruction

Conventionally, in the history of the White Rock site, much is made of its location, near a navigable river, easy access to coal, and engineered access to waterpower. Eventually, coal became doubly important as steam engines were introduced to support the water-powered infrastructure and eventually supplanted that waterpower.[1] Outside a few primary sources, the need for space to tip industrial waste is barely mentioned. Tipping is a vital part of the smelting process and a significant part of the White Rock story. Today's legacy landscape is derived from a remarkably dirty industry working for two centuries in a highly constrained geographical location. Aside from the industrial processes running on the site, the physical location focused the adverse effects of smoke and waste pollution on the surrounding landscape, devastating the land and its ecology. (Bridges and Morgan 1990: 277–81)

The second half of the eighteenth century saw a massive increase in copper production from White Rock and the other newly created sites in the valley. In 1809, the famous Hafod works opposite White Rock added massively to productivity, and inevitably, the pollution ended up on Kilvey and north beyond Llansamlet.

By 1802, the hill above White Rock was lifeless. The devastation was often a talking point for journalists and visitors. An 1812 diarist wrote: '…about a mile or two towards the entrance of Swansea, the appearance is frightful, the smoke of the copper furnaces having destroyed the herbage; and the vast banks of scoriae surrounding the works, together with the volumes of smoke arising from the numerous fires, gives the country a volcanic appearance.'[2]

The pollution on the dark and dead hillside was even sufficient to be mentioned by George Grant Francis in his 1881 celebration of the Swansea copper industry (Francis 1881: 115–16). This is possibly surprising as Francis was keen to talk up the industry in a fine example of boosterism, but it may reflect the unavoidably ugly state of the hillside as it appeared to tourists in the 1800s (Boorman 1986: 93). Later in his book, Francis returns to the subject describing Hafod being 'denuded of every vestige of verdure', with nature last 'having her sway' in the 1730s (Francis 1881: 139)

The death of nature in the shadow of industrialisation has been a theme famously explored by the critical feminist historian Carolyn Merchant (Carolyn Merchant 1980: 128).

Left: An extract from the industrial survey of South Wales (about 1905). This shows the peak industrialisation of the valley industry. After this point, the industrial giants of Germany and USA started to dominate technology and modern industry needed far larger extents for their industrial processes. It is interesting to speculate how much industrial waste still lies beneath the surface of the valley area.

Merchant sees some European industrial history as 'destroying' nature with seventeenth-century industry and technological expansion. She tracks how landowners (such as the Mansel family in Glamorgan) moved away from managing their land for wool or agriculture to new types of mining and metallurgical industries (furnaces, water supplies, and timber sources). The origins of status and wealth shifted from land to open-ended profit accumulation in the markets, with the status being purchased at the expense of nature as a resource base (Merchant 1980: 212–13). The concept certainly fits the Swansea copper industry experience. The gardens, pastures and fields of Cae Morfa Carw and Hafod disappeared under factories, and waste tips and the farms of Kilvey and Llansamlet were destroyed by atmospheric pollution.

The smoke from copper smelting was foul and heavily toxic, being one of the most potent forms of smoke pollution that emerged from the first industrial revolution. The suggestion that any such pollution was more than adequately compensated by the growth in Swansea's status and the money made is an awkward simplification (Miskell 2006: 71–72). In Swansea, as a centre of coal mining and coal use since the 1300s, coal smoke was undoubtedly a significant addition to the atmosphere. Still, the introduction of copper smelting propelled Swansea's pollution problem into another dimension. Introducing copper smelting technologies into the valley changed everything in the local environment. The chemical composition of the Cornish copper ore meant that once combined with local bituminous coal, smelting released vast quantities of sulphur dioxide, hydrogen fluoride, sulphurous and sulphuric acids, and hydrofluoric acid. Prevailing winds took most of this onto Kilvey and over Llansamlet and Winch-wen, although plenty of this pollution landed on the town when the wind changed direction. Toxic particles of copper, sulphur, arsenic, lead, antimony, and silver covered the east side. Particulates may have contributed more than anything else to the environmental destruction (Newell 1997). Quantitative data for the industry's early years doesn't exist, although we are left with one excellent source of insight into copper smelting in 1840s Swansea. The famous French sociologist Frederick Le Play spent several months between 1836 and 1846 examining every aspect of copper production at the Hafod works. Le Play's training as a metallurgist enabled him to ask the right questions as he studied the 'Welsh Method' of copper smelting and manufacture. This was published in a book with a wealth of detail (Le Play 1848). Modern historians have relied on Le Play extensively to summarise working conditions in the Hafod Works, particularly the nature of work for women and children (Evans and Miskell 2020: 95–99). But Le Play also looked at copper smoke pollution. His calculations provide a truly horrendous view of pollution. Le Play calculated that in 1848, Vivian's Hafod Works was releasing 188 tonnes of sulphuric acid into the Lower Swansea Valley every day (equating to over 65,000 cubic metres of vapour daily). He suggested that the wider Swansea Valley received 92,000 tons of sulphuric acid annually from all the valley copper works (Le Play 1848: 161–64). Le Play's calculations were even mentioned in George Grant Francis' celebratory history of the copper industry thirty years later (Francis 1881: 156–58).

By the early 1800s, slag tips quickly dominated White Rock and the riverside. A pair of

Above: Two images illustrating the extent of tipping in relation to the woodland on western Kilvey. The top image is a German Air Force image taken in preparation for air attacks in February 1941. By this time, the White Rock works had been shut and left to deteriorate for nearly twenty years. Many of the buildings have had slate removed and other buildings were in a dangerous state. The main tip is left of centre and owes its shape to the steam engine incline system that allowed waste to be moved up out of the White Rock and up onto the steeper slopes. There are no trees on this image. The lower image is a satellite composite showing the modern landscape. The current woodland is shown bright green. The purple areas are the extent of tips from the 1880s. The purple area far left is the slag pile on Cae Morfa Carw. From the 1830s, the works began tipping onto the hill over the original water power leat system. By the early 1900s, the brown area of much larger scale tipping was in use. The brown tip was being worked for building ballast in the 1940s, but the main removals took place 1965 to 1970 when material was moved to provide a base for Morganite. (QGIS composite images).

drawings from the end of the eighteenth century depicts the White Rock works surrounded by piles of industrial waste spilling into the river.[3] The drawings suggest that the waste tips were getting too big for dumping by wheelbarrow, and more 'industrial' measures were needed for tipping to continue. It's also hard to believe that the growing White Rock tips didn't threaten the navigation of the river, as the drawing shows unconsolidated waste rising to 4-5m above the river surface on the banks of the Tawe, a river renowned for its violent flooding and flow, and tendency to transport massive quantities of debris (and probably copper waste) into the Town Reach and Fabian's Bay. (Jones 1922: 158–67). The exact details of how waste was moved around the cramped White Rock site are not known to us; however, the issue as it affected the adjacent Vivian's Hafod works in the 1840s has been nicely summarised in some detail in recent work detailing the sheer hard manual labour of the early waste disposal part of the industry (Evans and Miskell 2020: 98–99). The lack of space within the works meant intensive work with wheelbarrows in the 1730s and, later, some rapid development of plateways for trams like those used in the coal industry. By the 1820s, the evolution of cableways and inclined planes alongside better wheels and tramrails meant that moving waste to the valley side became quite sophisticated. On the opposite riverbank, Vivian's Hafod works were experiencing similar problems with its smelting waste. By 1826, they installed a steam-powered winding engine to haul waste to the vast tip nearby at Pentre Hafod, which dominated the landscape until the 1960s (Hughes 2008: 30). There was eventually a similar arrangement at White Rock, with a tram line built up the side of the by then poisoned hill alongside the Nant Llwynheiernin valley, allowing successive layers of waste tipping onto the dead pasture. The inclined tramway went through several redevelopments as the problem grew. The steam winding engine and tracks were removed as part of the reforestation, but remnants of the inclined plane onto the hill still form part of the hill with odd concrete fragments on the surface and more visible remains within the White Rock Heritage Park (Hughes 2008: 23).

The growth of the tips on the Kilvey slopes raises a question about the change in perception of the importance of waterpower. The documents show that the need for waterpower for White Rock was a priority until the 1820s; after that, its importance quickly waned. Although there was a provision in the 1805 lease for the construction of a 'Fire Engine' (steam engine), this was to supplement existing waterpower provision through the Nant Llwynheiernin ("One Fire Engine for the purpose of supplying the Stream or Watercourse by which the Machinery of the said Copper and Brass Works hereby demised have been hitherto worked with a sufficient quantity of water for that purpose"), that being a pragmatic solution by pumping extra water into the existing channel.[4] Things changed by the time the OS 1:500 plan was compiled in the 1870s. By then, via the inclined tramway, tipping had spread over a large part of the lower reach of the Nant Llwynheiernin and started to fill in the original water channel. However, the drainage sluice to the original Knap Goch Millpond remained open (as it does today). The waterpower system that had evolved since mediaeval times and powered the first hundred years of copper production finally seemed to have lost its significance and value.[5]

Above: The incline from White Rock onto the hill. It is incredible to think that this small piece of land carried the turnpike road, two railway lines and a canal. This photo from 1932 shows its last years. In the foreground a small portion of the original landscape survives with the original course of Nant Llwynheiernin as it ran under the turnpike road bridge. The little wall to the right of the motor car is still there and the original drain of the Nant into the Knaploth grain mill is still underneath. The lower image is the 1:500 survey originally intended to cover public health issues but the sheet also included a snapshot of the slag tip as it expanded onto the hill in the 1880s. (Author's collection).

The development of inclined tramways and steam winding engines from the 1820s allowed freedom from the constraint of the manual barrow and small truck-based 'slag-tramming'. As a result, that constraint was removed at the moment of the most significant growth of the valley industries and the associated waste output (Miskell 2006: 80–86). Furthermore, adopting steam power and tram roads allowed for the wider ruinous distribution of waste across the hill.

Some sentiment about the loss experienced surfaces in the witness statements from the copper smoke trials of the 1830s when local farmers confronted the industry in a vain attempt to derive compensation or accountability from the big industry of the Vivians in Hafod (Rees 2000). The trials from the 1830s illustrate the immense difficulty we still see today in bringing polluting industries to justice. The proceedings recorded in impressive detail in the local newspaper display all the excesses of mendacious law practice and the power of industrial money. The Cambrian reported the evidence in an impressive amount, cataloguing the environmental catastrophe in detail that most people would easily recognise today ('Indictment of Messrs. Vivians' Copper Works…as a Public Nuisance' 1833). Still too early in Britain for the use of the word 'pollution' (Jarrige and Le Roux 2020: 88–90), the trials examined the role of smoke from the Hafod works in destroying the livelihoods of a group of farmers who farmed land downwind of Hafod. Nowadays, most people have some familiarity with the concept of pollution, perhaps as the degradation of an environment following the introduction of substances or activities that cause adverse changes to an ecosystem or environment (Jarrige and Le Roux 2020: 2–5). However, the destruction of Kilvey was linked to copper-smelting smoke because the smoke was the visual phenomenon most easily seen. The effect on groundwater and the ecology of the river and the bay must have been equally horrendous.

The proceedings of the one-day trial have been nicely summarised and discussed elsewhere (Rees 2000: 75–89). However, reading the original report from the Cambrian yields more in understanding the impacts of smoke on vegetation and animals. The effect on humans was not examined, and little evidence could have been given to illustrate any outcomes for public health. What little there was from two Swansea doctors was heavily caveated and fraught with the ignorance of the medical profession of the early 1800s in actually claiming health benefits from the smoke (Rees 2000: 86; 'Indictment of Messrs. Vivians' Copper Works…as a Public Nuisance' 1833: 4). The court report gives a distinctly unsavoury review of the legal process. The prosecution case rested heavily on the physical evidence of toxic pollution on the landscape and animals, with successive farmers relating tales of extreme malformations in cattle and horses and a collapse in the productivity in the hitherto productive farmland. The defence case relied on unpleasant attacks on the farmers both personally and professionally whilst pointing out that

Right: An extract from a rare coloured version of an image commonly used to show the forest of chimneys at White Rock in the 1860s The toe of the Cae Morfa Carw slag tip is left of centre and the mass of chimneys of Middle Bank is to the left. The drawing was commissioned by the French travel journal 'Le Tour du Monde ' and the artist was Jean-Baptiste Henri Durand-Brager (Le Tour du Monde) 1865) (Author's collection).

regardless of the environmental cost, the worth of the industry to Swansea, to the Vivians and the strategic safety of Great Britain was far more significant. The jury, no doubt mindful of the hefty social and political power of the Vivians, were extraordinarily compliant, returning a not guilty verdict for Vivian even before the Judge summed up the case.

The witness descriptions of the copper smoke are hard to envisage today in an age where pollutants are essentially odourless and colourless (although still pernicious). There are repeated descriptions of the smoke being dense black or blue and having a thick texture fouling the mouth and making breathing difficult. In wet weather, it was worse as the rain washed the filth from the air onto plants, animals, and people, destroying corn crops within hours. The filth hung over Kilvey and the east of the lower valley constantly. One witness described how in 1820, the Neath-Swansea mail coach was reduced to walking pace on the Llansamlet road with the guard forced to walk the horses forward with a lamp; such was the impenetrability of the murk. Considering that it happened in 1820, it is likely that such smoke events were near-constant daily occurrences from 1810 to at least the 1850s or even later, with relief only given by wind changes or stormy weather.

Across Kilvey, trees were a distant memory at best. Tree deaths must have started in the 1770s, with at least one artist unwittingly recording dead mature trees in his rendering of the picturesque White Rock works. Witnesses talk of oak, ash, and sycamore, all left as standing dead wood across Kilvey and Llansamlet.

The smoke affected animals in cruel and distressing ways. Forced to browse on the toxic grass, they quickly developed deformities of their mouths and teeth, with open sores exposing the bone. Hunger, emaciation, and death promptly followed. It was telling that the many horses working in the Vivian works were fed on food grown far west of Hafod. Hafod's managers knowing how to avoid the problems. A list of defunct farms was presented as evidence but quickly attacked by the mendacious Vivians' lawyer as evidence of poor agricultural practice and skills rather than smoke pollution.

Kilvey was described as 'barren as a road'; the fertile fields around the windmill at the summit that provided potatoes and barley and the 'good corn land' of the northern slopes were all disappearing or already dead. To reduce the community impact of the pollution, the White Rock operators had acquired most of the Kilvey farms east of White Rock by 1840, and the land was taken out of agricultural production. The lack of detail on the 1840s tithe plan of the parish suggests the surveyors, seeking any form of agricultural activity, found little interest in mapping there.[6] The dead land could have been considered fit for nothing better than dumping smelting waste. By the 1870s, the original Cae Morfa Carw marshes had been buried by a hectare of slag at least 4m in height and higher in parts. The riverside must have been physically

Top right: A sketch attributed to Philip James de Loutherberg from about 1786. Described as 'a large copper-smelting works' although it is plainly White Rock. The buildings are usually the focus but look at the piles of copper slag waste on Cae Morfa Carw and clearly tipping into the river. The artist has also hinted at the masses of black smoke produced by all those chimneys. Tate Gallery © Tate, 2019.
Lower right: Part of the slag field of Cae Morfa Carw was exposed in 2007 allowing a view of the slag that made up the tips from the above drawing. (Author's collection).

incapable of taking any more waste by the 1820s, and it does not experience much change in successive OS maps and surveys until the 1960s. Tipping started on the hill's slopes in the 1820s, just as the need for waterpower began giving way to steam engines.

By 1879, the extent of the 'secondary tip' east across the turnpike and up to Cnap Goch House had spread to at least 5 ha (12 acres) and some considerable height, needing a tramway to dump the slag. The 1870s OS 1:500 plan shows horrifying extents of slag and waste in octopus-like tentacle shapes, reflecting the meandering of tram lines dumping successive layers of waste. Via the inclined tramway, slag was spread up to and beyond the old house, which survived as a ruin surrounded by waste until the 1880s. The tip's activities depended on the fortunes of the White Rock Works, but by 1924 the works had closed, with tipping ending shortly after. By that time, the tip had expanded to about 12 ha (29 acres) (Robins 1993: 6). The tip continued to be worked and extracted as suitable ballast for roads and embankments until 1967 (Borough Engineer, Surveyor and Planning Officer's Department 1969: 2). Famously, a lot of the northern tip slag was used to raise the land surface above the Tawe's flood plain for the Morganite site at the north end of the Enterprise Zone (Bromley and Humphrys 1979: 186–87).

Recent figures suggest that each tonne of copper in the 1730s could have generated approximately twenty-six tonnes of waste. However, I constantly wonder if that is underestimated (Symons 2003: 92). The early leases have substantial clauses ensuring space for dumping waste. This is where the physical geography of the site becomes of crucial importance. The original White Rock site, as leased in 1735, was barely more than 2.4 hectares (6 acres), which left about two hectares (c. 4.5 acres) of adjoining fields and saltmarsh (known as Cae Morfa Carw) for tipping of smelting waste. The fields connected a larger enclosure to the north (Morfa Carw), which the Duke of Beaufort owned. This northern enclosure would eventually become Middle Bank Works. Much of the land was just above sea level, and the land adjoining the river was stone shoals, salt marsh, and pasture. Nowadays, most of the urban reach of the river has been embanked or hidden under a barrage. However, a flavour of the original character of the river where it passed Morfa Carw and White Rock can be gained by looking upriver to the river bed at the 2014 flood scheme near Ynys Forgan

A 1771 plan of the river gives the best indication of the vast extent of salt marsh along the riversides of both Hafod and Morfa Carw (Jones 1922: 55). Although the 1730s river was described as 'navigable' upstream to Morfa Quay, the usable channel must have been extremely narrow and subject to disruption by the vagaries of the river. The 1771 channel looks no more than 10m wide in places. The 1744 drawing mentioned earlier provides a vivid impression of the salt marsh fringing the river with grazing animals and the meadows behind it. We can see the grassland and hedgerows being steadily submerged beneath the advancing piles of tipped ash and slag from the copper works side. The artist has depicted a series of dead trees, likely casualties of the early pollution.

The land surface we have today at White Rock is about 4m higher than the fields of Morfa Carw, but only because the landscaping of the 1990s levelled out a lot of the small mountain of waste that marked the location of the original Cae Morfa Carw. The central hill of the White Rock Heritage Park is a lasting gravestone of industrial waste over that ancient meadow.

Even though the early leases made provision for using the Morfa Carw meadows as a tip, all parties knew that would not be enough. There was a concern that slag would end up tipping into the river or spreading out from the tips on the meadows, which would have limited that narrow navigable channel. To prevent that, the leases had rather weakly worded clauses to prevent slag and ash from being thrown into the river to "prejudice the navigation of the River Tay".[7] Over the years, this seems to have had limited effect, and a sketch of White Rock from the end of the 1700s clearly shows waste tips had engulfed the whole of Cae Morfa Carw and were spilling into the river.[8] The other insurance offered was to allow tipping waste on adjoining Mansel Estate lands which can only have meant the land around Cnap Goch House north of the White Rock site. The road to that tipping site was well established as part of the Turnpike network, and Ty Knap Coch (Cnap Goch House) and its surrounding land were prominent on the Lightfoot drawing in 1744, the road being an essential route for coal traffic coming from the Mansel Estate down to the Mansel Coal Yard at White Rock Quay. As this was quite an uphill slope, it is likely that tipping there was a choice only taken when space on Cae Morfa Carw had run out, possibly by the early 1800s. By the 1820s, most Lower Swansea Valley industries were building inclined railways onto new tipping sites to deal with the mountains of waste from their smelters (Hughes 2008: 30)

The polluting phase of White Rock lasted just over 188 years. The other metal industries of Hafod and Landore would grow and eventually dwarf White Rock in both the profits made and their negative environmental impact (Bromley and Humphrys 1979: 2, 12). Slag from the valley became a near-ubiquitous Swansea building and repair material. In the early years of the nineteenth century, stone walls throughout the town were regularly patched with slag blocks instead of Pennant Sandstone. I recall coming across a slag-based repair to the eighteenth-century Cadle corn mill leat in the heart of the Llan valley, several kilometres from the copper works, even though local stone was readily available at Glyn Silling nearby. Much is made in heritage terms of the slag coping stones of the Trevivian streets (Hughes 2008: 53). But it is reasonable to accept that Swansea was far more than just a copper town, as has been ably discussed elsewhere (Miskell 2006: 77–73). Swansea was wrought by coal mining, with lasting impacts above and below ground. The urban areas are fringed with coal tips, and remnants of the mining industry, particularly from Roman times until the 1820s, are often only ever found when construction necessitates excavation. The slag tips are less recognised in the various histories of Swansea's industry.[9] Their ability to poison land over successive generations often prompted their removal or levelling. When the Great Western Railway built their railway into Swansea across to Landore, the engineers relied on terraces of glacial gravel as embankments (Strahan 1907: 134–35), the later embankments for the northern part of the Enterprise Zone and the M4 relied on massive quantities of metalliferous slag. But just as the industry had a profound effect on Swansea's history, the pollution left behind had (and still has) a profound effect on Swansea's geography.

Notes

1. ('Copy of Lease Dated 21 September 1805' 1805: 14) See for example the clause inserted into this lease to allow the construction of a steam engine to pump more water into the Knap Goch leat or Nant Llwynheiernin to supply the White Rock Battery.

2. D. Webb, 1812 as quoted in Newell (Newell 1997).

3. (Philip James De Loutherbourg [n.d.]).

4. ('Copy of Lease Dated 21 September 1805' 1805: 14).

5. (Ordnance Survey of Wales 1879) This series of plans at 1:500 scale covers the urban area of the town in the 1870s. Luckily, coverage extended to include mapping of the White Rock works and its waste tip.

6. ('Welsh Tithe Maps - Home' [n.d.]) The Apportionment of the Rent-Charge in lieu of Tithes in the Parish of Lansamlet, White Rock Works, p.14.

7. ('Abstract of Lease Dated 21 March 1736' 1736: 7).

8. (Tate 2019)

9. (Williams 1940; Evans and Miskell 2020) Are good examples of influential works.

REFERENCES

'Abstract of Lease Dated 21 March 1736'. 1736. , West Glamorgan Archives Service, DD Xhr 31

Boorman, David. 1986. The Brighton of Wales: Swansea as a Fashionable Seaside Resort, c.1780-1830 (Swansea: Swansea Little Theatre Company)

Borough Engineer, Surveyor and Planning Officer's Department. 1969. A Report of the Reclamation of the Former Tip Complex of the White Rock Copper Works, Swansea, West Glamorgan Archive Service, PL11/17/7

Bridges, E.M, and Huw Morgan. 1990. 'Dereliction and Pollution', in The City of Swansea: Challenges & Change, ed. by Ralph A. Griffiths (Stroud: Alan Sutton), pp. 270–90

Bromley, Rosemary D. F., and Graham Humphrys (eds.). 1979. Dealing With Dereliction: The Redevelopment of the Lower Swansea Valley (Swansea: University College of Swansea)

'Copy of Lease Dated 21 September 1805'. 1805. , West Glamorgan Archives Service, DD Xhr 32 [accessed 23 September 2021]

Evans, Chris, and Louise Miskell. 2020. Swansea Copper; A Global History (Baltimore: Johns Hopkins University Press)

Francis, George Grant. 1881. The Smelting of Copper in the Swansea District of South Wales from the Time of Elizabeth to the Present Day, 2nd edn (London: Henry Southeran)

Hughes, Stephen. 2008. Copperopolis: Landscapes of the Early Modern Industrial Period in Swansea (Aberystwyth: Royal Commission on the Ancient and Historical Monuments of Wales.)

'Indictment of Messrs. Vivians' Copper Works... as a Public Nuisance'. 1833. The Cambrian (Swansea), pp. 3–4

Jarrige, Francois, and Thomas Le Roux. 2020. The Contamination of the Earth: A History of Pollutions in the Industrial Age, History for a Sustainable Future (Cambridge, Massachusetts: MIT Press)

Jones, William H. 1922. History of the Port of Swansea (Carmarthen: W. Spurrel and Son)

Le Play, Frédéric. 1848. Description des Procedes Metallurgiques Employees dans le Pays de Galles pour la Fabrication du Civre (Paris)

'Le Tour Du Monde : Nouveau Journal Des Voyages / Publié Sous La Direction de M. Édouard Charton et Illustré Par Nos plus Célèbres Artistes'. 1865. Gallica <https://gallica.bnf.fr/ark:/12148/bpt6k343878> [accessed 16 March 2023]

Merchant, Carolyn. 1980. The Death of Nature: Women, Ecology and the Scientific Revolution (New York: Harper and Row)

Miskell, Louise. 2006. 'Intelligent Town': An Urban History of Swansea, 1780-1855, Studies in Welsh History (Cardiff: University of Wales Press)

Newell, Edmund. 1997. 'Atmospheric Pollution and the British Copper Industry, 1690-1920', Technology and Culture, 38: 655–89

Ordnance Survey of Wales. 1879. 'Glamorgan XXIV.1.14' (London: Ordnance Survey) <http://hdl.handle.net/10107/4715288>

Philip James De Loutherbourg. [n.d.]. A Large Copper-Smelting Works at Swansea (Perhaps the Copper Works at Landore on the River Tawe, North of Swansea) <https://www.tate.org.uk/art/artworks/de-loutherbourg-a-large-copper-smelting-works-at-swansea-perhaps-the-copper-works-at-d36374>

Rees, Ronald. 2000. King Copper: South Wales and the Copper Trade 1584-1895 (Cardiff: University of Wales Press)

Robins, Nigel A. 1993. Eye of the Eagle: The Luftwaffe Aerial Photographs of Swansea (Swansea: Tawe History)

Strahan, Aubrey. 1907. The Geology of the South Wales Coal-Field. Part VIII, The Country around Swansea: Being an Account of the Region Comprised in Sheet 247 of the Map, Memoirs of the Geological Survey. England and Wales, 247 (London: Printed for His Majesty's Stationery Office by Wyman and Sons)

Symons, John C. 2003. 'The Mining and Smelting of Copper in England and Wales, 1760-1820' (unpublished masters, Coventry University in collaboration with University College Worcester) <https://doi.org/10/10._Bibliography.pdf>

Tate. 2019. '"A Large Copper-Smelting Works at Swansea (Perhaps the Copper Works at Landore on the River Tawe, North of Swansea)", Philip James De Loutherbourg, 1786 or 1800', Tate <https://www.tate.org.uk/art/artworks/de-loutherbourg-a-large-copper-smelting-works-at-swansea-perhaps-the-copper-works-at-d36406> [accessed 27 November 2021]

'Welsh Tithe Maps - Home'. [n.d.]. <https://places.library.wales/viewer/4570048> [accessed 27 November 2021]

Williams, D. Trevor. 1940. The Economic Development of Swansea and of the Swansea District to 1921, Social and Economic Survey of Swansea and District, 4 (Swansea: University of Wales Press Board)

Five: Repair

Even in a town that can lay claim to being one of the world's earliest industrial areas, industrial heritage has often been a challenging sell. With industrial archaeology existing throughout so much of the subsurface of the urban area, the unplanned and often chaotic building of industrial buildings, canals, railways, and particularly waste disposal throughout Swansea's early industrial revolution left a legacy of poorly documented sub-surface problems which remain and can always threaten any structure or redevelopment scheme (Director of Environment 2008: 4–5).

Celebrating Swansea's industrial past requires great imagination and artistic interpretation, as most industrial buildings were demolished over the past century. Those that survived have been heavily altered or rebuilt after decades of neglect at considerable new expense.

Despite Kilvey's industrial devastation and the upheavals of the post-industrial phase from the 1960s, there are remarkable survivals of the eighteenth-century landscape which have escaped notice over the past decades. This is understandable; the star of the show was often the White Rock ruins or even the regret over their loss, because they were mainly destroyed between 1963 and 1965 (Borough Engineer, Surveyor and Planning Officer's Department 1969). The complex phases of building, new technologies and rebuilding resulted in a complicated layout of buildings as compromises were made to suit industrial processes to the small site. The ruins that survived into the 1940s were so confusing that in 1941, the Luftwaffe thought they had bombed the area and claimed it as successful bomb damage.[1] The early 1970s were a frenzy of remediation and tree planting to hide the industrial scars. The 1990s bridge and road junction remodelling of Middle Bank, north of the White Rock site, finally removed many surface features that survived. The area is due for more upheaval to accommodate the new tourism plans for a foreign investment company. However, that is far from the end of the story; the subsequent history of environmental restoration and repair may be the best story of all.

'...a giant rubbish dump of tips on either side of the river which had become an industrial sewer,' (Hilton 1967: 1)

So began the famous Lower Swansea Valley Project's initial justification in the introduction to the project summary (known as the Hilton volume) in 1967. The 'rubbish dump' was massive,

Left: Weathered rock exposures from the eighteenth-century along the banks of Nant Llwwnheiernin above Martin's Pond. This sandstone shows the results of two centuries of exposure to acid rain and copper smoke from the White Rock works. This part of the hill would receive over 180 tons of sulphuric acid every day between the 1760s and 1890. The exposure is directly in line with the original location of the early chimneys of the works. (Author's collection).

with at least 405 ha (1000 acres) of land choked with toxic waste from two centuries of industrial tipping. The toxic characteristics of the waste meant that, although it wasn't the largest body of derelict land in Wales, that dubious honour was held by the coal tips of the Glamorgan valleys; it was undoubtedly the most toxic and troublesome (Bromley and Humphrys 1979: 14–18). However, with a characteristic bluntness from a man with a reputation for getting to the point, His Royal Highness the Duke of Edinburgh, in the preface to the Hilton volume, was more direct, describing the valley as 'a stark monument to thoughtless and ruthless exploitation. While it remains in its present state, it is a standing reproach to each generation which shrugs its shoulders and looks the other way.'

'Pollution' as a term can be as problematic as 'nature' (Gandy 2020: 620–21). Although we can understand and accept that pollution degrades our environment today, we can see that on a European scale, the concept has taken at least two centuries to appear as a common understanding (Jarrige and Le Roux 2020: 5–6). 'Pollution' evolved gradually out of the legal category of 'nuisance' throughout the nineteenth century (Jarrige and Le Roux 2020: 5). In Swansea, the talk was of 'Copper Smoke' as the most visible sign of industrial exploitation. Reading the history of Swansea's legal actions over the 'nuisance' of copper smoke in the 1830s makes for depressing reading (Rees 2000). The lasting impression is of a town in thrall to the millionaire industrialists and their talented legal representation intent on diminishing or dispelling any impact of the copper industry on the environment. These 'merchants of doubt' tackled farmers' complaints of destroyed land and livelihoods by any means possible.

By the 1850s, the White Rock area was described as having 'devolved into an industrial oven', such was the extent of the smokestack forest between Hafod, Landore, White Rock and Middle Bank (Jarrige and Le Roux 2020: 179–81). The legacy of pollution stays horrible. Swansea is easily identified on the national geochemical atlas of Wales by the acute hotspots of copper, cadmium, and lead that still afflict Kilvey and the valley (Webb and others 1978: 27, 41). The thin veneer of coniferous woodland over the western side of Kilvey covers a layer of toxins easily disturbed and always capable of wreaking untold damage on plant health (Broad 1979: 12, 28). Regarding people, very little medical research was ever conducted on the health implications of the waste tips and air pollution, although it was hinted that there were some links between coronary heart disease and heavy metal pollution in West Glamorgan in the 1960s (Bromley and Humphrys 1979: 92–93)

The Lower Swansea Valley Project provided funding and expertise in the early 1960s to address the legacies of air pollution and waste tips (Lavender 1981: 5). Although clearances and tip removals continued into the early 2000s, the main clearance phases were finished by the late 1970s with a milestone conference held in the University College of Swansea that celebrated the remarkable multidisciplinary partnership between the local university and local government and the Welsh Development Agency (Bromley and Humphrys 1979). The work was considered unique in Britain, particularly in terms of engagement and community involvement (Broad 1979: 11). Ironically, the Kilvey tip area was not included within the scope of the Lower Swansea Valley Project. Instead, Kilvey was seen as an aesthetic landscape problem rather than part of

Above: a newspaper photograph of the Gasworks Nursery in the late 1960s showing schoolchildren looking at seedlings being prepared for replanting as part of the Lower Swansea Valley Project. (Author's collection).

the valley's human and economic challenges, which required better employment, transport, and housing facilities (Bromley and Humphrys 1979: 15). In the early 1960s, pioneering investigation of the valley's waste tips enabled a thorough understanding of the chemical and structural nature of the tips and their complex interrelationships (Holt and others 1966). The valley and hillside had waste from coal tips, iron and steel works slag, non-ferrous works slag, and rubble from countless ruined buildings (Hilton 1967: 55–70). Just outside the project boundary, Kilvey Hill had at least 12 ha (about 30 acres) of waste from White Rock and enormous piles of rubble and even coal tips dating back at least 500 years. Added to this was at least 100 ha of destroyed or eroded topsoil from centuries of smoke. Whereas in earlier times, waste was often tipped by wheelbarrows and was small in granular size (with lumps of 2-3cm), this was not the case with later slags originating from more advanced smelting technologies, which produced far larger chunks of material. The building of the inclined railway up the side of the hill allowed for the tipping of hot slag direct from furnaces. As it cooled, this slag often clumped together in larger lumps which were very dense, heavy and much harder to deal with (Borough Engineer, Surveyor and Planning Officer's Department 1969: 2). Immense multi-coloured moonscapes of dead land and rubble greeted a traveller entering Swansea via the main railway. Travelling by train into Swansea via Brunel's famous viaduct gave a helicopter-style view of the vast expanses of poisoned land.

The chaotic style of industrial growth in nineteenth-century Swansea had, by the 1960s, left a legacy of complex and complicated leasehold and freehold land ownership in the valley and on Kilvey. Land on Kilvey had been acquired by the various operators of White Rock from the early 1800s for tipping of waste and to avoid compensation to farmers and landowners for the poisoned land. In the 1960s and 1970s, the Swansea Council Estates Department, often the unsung heroes of land restoration and repair of land, worked hard to bring the chaos into ordered public ownership (Bromley and Humphrys 1979: 11). A process not actually completed until 2022, when final applications for possessory title of land at the top of the hill still missing its original title deeds were completed, awarding title to Swansea Council in advance of the new redevelopment plans for adventure tourism. Dealing with the land assembly and ownership of the industrial valley was always going to be complex, but Kilvey proved equally challenging as many deeds and leases had been destroyed or lost, making historical enquiry difficult as well as future management. By 1969, the whole Lower Swansea Valley Project area, including Kilvey adjoining, had been recognised, categorised, and labelled. Of the 70 'plots' identified across the valley, the 'mountain' of Kilvey became known as Plot 57 (Bromley and Humphrys 1979: 186–89). Brutally bureaucratic possibly, but Kilvey, for the first time in decades, had an identity beyond the ignominy of being the 'White Rock Tip'.

The newly designated Plot 57 now had an identity and a spatial extent, and work could begin to address the devastation. By late 1969, Plot 57 was jointly managed by Swansea Council and the Forestry Commission (Bromley and Humphrys 1979: 187). Restoration could begin.

Plot 57 now had the benefit of consideration about how it could be valued environmentally for the first time in centuries. The relationship

Above: Contractors spraying grass seed and phosphatic fertiliser above Steam Engine Spring in Kilvey Woods in the summer of 1970. In the background is the newly recreated lower valley of Nant Llwyn-heiernin. These slopes were ploughed and limed in early 1970 and this photo was taken before the 'Gasworks Pines' had been planted. The Forestry Commission had considerable experience of restoration on South Wales coal tips and this was a scene very common across the former mining areas of Glamorgan in the 1960s. Kilvey was regarded as a 'severely degraded' soil and supported very little vegetation apart from isolated stands of Purple Moor-grass (Molinia caerulea). Preliminary liming of the ground was an attempt to fix the pollutants in the soil. The slopes were ploughed whenever possible although the steeper gradients of Kilvey were not touched.

The Commission relied on pines on Kilvey, with compartments of Scots, Lodgepole and Corsican planted across the hill. Tree death was commonplace in the first years of planting and regular re-planting was needed in the early years. The first year's plantings in 1970 were almost completely killed due to emissions from the Imperial Smelting Corporation's works at Llansamlet, and tree health did not improve until the works closed in 1974.

Although the ploughing was considered beneficial for tree health, there was no evidence to support a view that liming and adding phosphates to the soil had any beneficial effect. (Author's collection).

between the land and the town could be re-evaluated, and remediation, restoration, ecosystems and recreation could now be considered. The role of the central valley (Nant Llywnheiernin) was last considered necessary in the early 1800s as a source of industrial waterpower. It now had a new significance in, amenity and nature maybe even ecology. Removing tip waste and restoring the central valley of the Nant Llywnheiernin was a necessity. Centuries of coal mining had irretrievably diminished the surging water of the original stream. However, the upper portion of the valley still suggests the impressive size of the original stream from the 1600s as well as early attempts to work the Kilvey coal veins that run along the north bank. Removal of tip material throughout 1969 revealed the true extent of the damage with destroyed soil, toxic debris, craters, and scars left by bulldozers. The heavy equipment proved to be a benefit in moving the massive lumps of slag and even many of the substantial glacial boulders that littered the hill. By the end of 1969, the Council had created a remarkably faithful reinstatement of the original stream valley (Borough Engineer, Surveyor and Planning Officer's Department 1969: 3–4).

The moonscape that remained must have been quite daunting. Although the tip spoil was removed, the underlying soil had disappeared, leaving nothing to work with. Natural regeneration was never going to happen. The Forestry Commission believed that the exposed seaward slope would never produce more than patchy scrub (although recovery in the 1990s proved more optimistic) (Broad 1979: 26–27). This is where the knowledge and capabilities of the Lower Swansea Valley Project proved of broader benefit. The early 1960s re-vegetation research of the Project now supplied priceless advice confirming earlier theories and experiments on re-vegetating industrial sites elsewhere in Europe. More importantly, the Project provided practical experience, knowledge, and suitable tree species that may survive on the blighted land (Hilton 1967: 72–88). In a tentative approach to trying something that would work, Swansea Council Parks Department planted 200 trees in 11 small zones on either side of the newly recreated Nant Llywnheiernin . The trees, Scots Pine (*Pinus sylvestris*) and Monterey Pine (*Pinus radiata*), were specially grown at the Project's small nursery in nearby Morriston, the 'Gasworks Nursery') (Borough Engineer, Surveyor and Planning Officer's Department 1969: 4–5). These two species were renowned elsewhere for flourishing on impoverished soils, with Monterey Pine being a favourite for Welsh mining tips. The 200 trees were planted with some fertiliser and optimism in the autumn of 1968. The first time in over two centuries that trees had been planted on the hill.[2]

Plot 57 was going to need a lot more support to recover. Those first 200 'gasworks' trees were supplemented by work to introduce basic grass and clover vegetation (Borough Engineer, Surveyor and Planning Officer's Department 1969: 4). A mixture of seeds and fertiliser was sprayed across Plot 57 in conjunction with Forestry Commission ploughing in preparation for commercial planting of various tree species the Commission had learned would work well

Right: Although no longer running with water, the eighteenth-century leats above White Rock retain moisture within the dry coniferous woodlands and act as biodiversity hotspots throughout the forest. (Author's collection).

on the bleak devastation of opencast mining operations across Glamorgan (Broad 1979: 5). This was not exactly 'habitat re-creation'; conservation seed mixes were still in their experimental phases in the early 1970s. Initially, the plan was for Kilvey to produce a commercial timber crop (Gilbert and Anderson 1998: 47–48; Broad 1979: 10). A second priority was making the hill look green, an experience for the town unknown since the 1730s and establishing some vegetation to prevent further erosion and runoff. This 'kick-starting' of the succession would take three decades to make meaningful progress (Gilbert and Anderson 1998: 3).

Despite a certain unease about planting in a semi-urban setting with risks for public access, theft and fires, the Forestry Commission, in their usual practice, set up twenty-eight management compartments across Plot 57. After deep ploughing of some of the hillside, they began planting areas of Lodgepole Pine, Larch, and Scots Pine. The ploughing supposedly removed what remained of the surface landscape history, such as the pre-industrial roads and hedgebanks. However, as mentioned earlier, a surprising number of features relating to the industrial history of Kilvey survived on the steeper slopes where ploughing was impossible.

Reviewing the existing documentation covering the first few years of the reforestation of Kilvey, it is easy to see the nervousness of the Forestry Commission in establishing commercial woodland in South Wales. Locally, Commission plantation work began in the 1920s on the coastal hills above Port Talbot and Margam. The unique Glamorgan landscape of linear valley settlements, open moorland ranges separating the valleys, and chronic coal and metals pollution pushed the Commission into pioneering what we would be happy to call 'urban woodland' today. For decades, the corporate 'memory' of Forestry Commission Wales was clouded by the extreme difficulties of managing the Margam Forest lying 16km east along the coast. Margam's forest was established after the First World War as part of the Forestry Commission's mission to renovate Britain's dismal timber resources (South (Wales) Conservancy 1952). Margam's location seemed ideal, with proximity to the coal mines that would provide a significant market for pit props, even though the Great Western Railway was already committing substantial resources to build over 16 ha (about 40 acres) of timber ponds for imports at Barry Docks (Appleby 1933: 47–55). The Commission were under pressure to quickly provide pit wood for the coal mines and rushed into planting with species generally regarded as quickly productive elsewhere. (Rackham 2015: 357) The species chosen were Douglas Fir (*Pseudotsuga menziesii*) and European Larch (*Larix decidua*). The successive plantings in the 1920s and 1930s were disastrous, with subsequent official reports describing the work as 'distressful', 'wasteful', and widespread failures (South (Wales) Conservancy 1952: 30–40). Although the Margam saga is long and educational, apart from unwise species selection, two issues emerge that, whilst not uniquely Welsh, are nevertheless distinctively local: fire and air pollution. Margam's local farmers were slow to adapt to the fire risks of the new plantations and continued the regular maintenance burnings of their rough grazing (South (Wales) Conservancy 1952: 5). Fires inevitably jumped into the young trees, destroying hectares of new planting almost annually. The 'great fires' of March 1929 killed over 306 ha (750 acres) of newly planted

Above: Badger activity on the eighteenth-century coal tips of Kilvey in 2021. (Author's collection).

forest. The event endured in the memory of the Commission, who referred to the event nearly twenty five years later in an appraisal of 'progress'. The 'Great Fire' was followed by many others, including significant losses in 1936, 1942, and 1949. By the 1950s, Margam had distinguished itself as the Forestry Commission's worst fire risk. By 1952, over 600 ha of the original 900 ha of the plantation had been replanted at least once because of fire destruction (South (Wales) Conservancy 1952: 4–8). Lessons on fire defence, species selection, seedling size, planting distances and bitter experience would be applied to Plot 57 planting.

Understandably, the Forestry Commission viewed the forestation of Plot 57 as a commercial exercise; it was, after all, the raison d'etre of the Commission. The one attempt at some amenity plantings were the 'gasworks pines' mentioned earlier., Although a small exercise in visual amenity, it has left Kilvey with some glorious significant little pine trees. At that time, the concept of 'nature conservation was not yet in the lexicon of the Commission or the Council, probably understandable as in 1970, much of the valley and Kilvey looked beyond repair; how could anything be restored or recreated? And corporate bodies and local government were notoriously slow to learn. Even as late as 2004, some influential British authors complained about the intolerance for urban or industrial nature (Dunnett and Hitchmough 2004: 611–12). In 1970, Britain was still four years away from Bunny Teagle's landmark study of urban wildlife in Birmingham (Teagle 1977). Teagle's work has a special place in the history of nature conservation as he challenged the often-common phrase in municipal planning: 'there was nothing there'. Teagle studied the surviving nature of the Black Country, a place that readily displayed the brutal consequences of industrial blight in much the same way as Swansea (Goode 2014: 12). Teagle's work destroyed the myth of the urban wildlife desert; it may even have changed the rules of nature conservation overnight. Teagle's original work still reads well. The detailed appendix listing the plants and animals he discussed is still exciting (Teagle 1977: 20,53). Teagle's work also helped better focus a century of British urban natural history and research from the local scientific and field naturalists' societies that had escaped mainstream attention. It provided a wealth of botanic and geographical information from pioneering enthusiasts on regular field trips using the dense railway network of Edwardian times or enjoying the new charabancs and buses in the 1920s. Locally, the Swansea Scientific and Field Naturalists Society was very active between 1919 and 1947 and provided a wealth of information about local plant and animal presence, including some interesting research on industrial nature (Pugsley 1941). Beyond trees, the 1970s work on Plot 57 had very little to do with habitat restoration, and a reader will search in vain for nature conservation in the three main books on the restoration of the valley (Hilton 1967; Lavender 1981; Bromley and Humphrys 1979).

The literal 'baptism by fire' the Forestry Commission underwent in Margam eventually benefited Swansea. It created considerable internal capability and experience for managing new woodlands on Welsh post-industrial sites. By 1980, the Commission had an impressive forestry portfolio on opencast coal tips, colliery spoil heaps, degraded industrial soils, old quarries and derelict wastelands in Swansea. Kilvey's Plot 57 was one of that portfolio's only two copper waste sites. The other is 'Copper Mountain' at Cwmafan, twelve kilometres east in the Afan Valley. The Cwmafan site was a Rio Tinto works that operated up to 1912 and poisoned the surrounding mountains, notably 'Copper Mountain' (Foel Fynyddau) next to the works (Broad 1979: 27–28)

From 1970, the Commission spent over 15 years planting pine and larch across Plot 57 in a series of compartments. Commercial assessments of the timber yield from the hill were not promising, reflecting the poor condition of the soil or soil remnants. Nevertheless, regular planting and replanting ('beating-up in forestry jargon) continued, much of which the Commission carefully recorded in their planting records.[3] The tree growth, slow to start and with many failures and constant fire damage, started to look a lot greener. By the mid-1980s, sufficient tree growth and change meant renaming Plot 57 to 'Kilvey Woods' was possible.

If the 1980s saw much change on the hill, this was also the case for the Forestry Commission. International markets for timber changed beyond all recognition and the rigid civil-service style production policies that were the 'comfort zone' for British forestry vanished. In Wales, local authorities, forestry workers, tourists and forest users increasingly challenged older industrial afforestation techniques (Owens 2009: 178–82). For example, the 1970s surge of interest in the Swedish sport of Orienteering encouraged the Commission to cater to new types of users. Orienteering relied on the comprehensive mapping of many Welsh forests at scales and levels of detail far beyond the services of the Ordnance Survey. Not surprisingly, users became far more aware of the woodlands around them.

Equally, newly-trained forest managers brought more modern techniques from their colleges, which directly challenged the conservative capabilities of the Commission. Benefits were no longer considered as purely economic. The roles of wildlife, landscape, and recreation forced the Commission to slowly move to 'multi-purpose' woodlands that addressed wider stakeholder needs. Progress was slow, and viewpoints only changed as older staff retired and recruits arrived with modern ideas. The concept of 'post-industrial' forestry and woodlands emerged in the late 1980s, increasing recognition of the need for wider stakeholders and benefits in our forests (Mather 1991). For Kilvey, the change in perception from timber resources to wildlife reservoirs and habitats started in 1988 when the local authority began to warm to better environmental policies. The key drivers were changing attitudes within the County, and Swansea City politics, often from closer liaisons with the local university as academic fashions evolved towards more recognition of environmental issues and how the environment was managed in some European countries (Lachmund 2017). There was also wider acceptance politically that it wasn't all about tourism on the Gower peninsula and that the urban nature of Townhill and Kilvey needed to be better recognised and managed (City of Swansea 1989). To a certain extent, the political changes managed to reinvigorate the incredible momentum of the Lower Swansea Valley Project (Griffiths 1990: 293–99). The legal business of creating a transition from industrial forest to community woodland followed in 1991 (Wilmot and Harris 2009: 4–5). From the mid-1990s, a series of Forestry Commission (Wales) funded projects began to address issues of access and education, not least trying to address the chronic fire setting and fly-tipping that scarred the woods.

The quest for economic rejuvenation of the valley area began in earnest in the early 2000s, building optimistically on the resurgence of interest created by two notable books about Swansea's copper industry (Hughes 2008; Rees 2000). Any modern re-evaluation of the industrial archaeology of the site now needed to include the wider environment, not least to extend the scope of the hunt for funding as wide as possible. In an insightful section, the 2012 re-evaluation of the surviving Hafod ruins and their reinterpretation as a heritage park acknowledged the role of the 'icon' of Kilvey Hill and its regeneration as part of the industrial heritage of the wider area (Goskar 2012: 14). Despite some 'critical ambition' with an eye on UNESCO and world heritage status, the extent of 1960s site clearance in Swansea has probably ruled out the establishment of a World Heritage Site similar to Blaenavon (Reynolds and Gibson 2015: 27). Equally, it would be more appropriate to commemorate the environmental recovery from the industrial devastation. However, the green core of the Lower Swansea Valley and Kilvey will likely be of more local benefit to biodiversity, environment, and community than the bleaker landscapes of the heritage iron or slate industries.[4]

Notes

1. See Robins, 1993, photo 6, dated 15 February 1941. A number of derelict buildings were identified as bomb-damaged by German air force photo interpreters because their roofs were missing. However, the buildings had been ruined and left derelict twenty years earlier.

2. In spring 2022 I went in search of these pioneer trees. Incredibly, over 100 had survived with the Monterey Pines doing particularly well and now with trunk diameters of c. 80cm enjoying the distinction of being the largest and oldest trees on the hill.

3. See for example, the Forestry Commision/NRW sub-compartment data (lle.gov.wales) Accessed February 2022.

4. (Barber 2002) A beautifully readable tour through the Blaenavon World Heritage Site.

REFERENCES

Appleby, H.N. (ed.). 1933. Great Western Ports (Cardiff: H. N. Appleby)

Barber, Chris. 2002. Exploring Blaenavon Industrial Landscape World Heritage Site (Abergavenny: Blorenge Books)

Borough Engineer, Surveyor and Planning Officer's Department. 1969. A Report of the Reclamation of the Former Tip Complex of the White Rock Copper Works, Swansea, West Glamorgan Archive Service, PL11/17/7

Broad, K.F. 1979. Tree Planting on Man-Made Sites in Wales (Forestry Commission) <https://www.forestresearch.gov.uk/documents/6846/FCOP003.pdf> [accessed 23 September 2021]

Bromley, Rosemary D. F., and Graham Humphrys (eds.). 1979. Dealing With Dereliction: The Redevelopment of the Lower Swansea Valley (Swansea: University College of Swansea)

City of Swansea. 1989. A Strategy for Greening the City (Swansea)

Director of Environment. 2008. The Environment Act 1995: Part IV Local Air Quality Management Review and Assessment of Air Quality City & County of Swansea Progress Report 2008 May 2008 (Swansea: City and county of Swansea) <https://uk-air.defra.gov.uk/assets/documents/no2ten/Local_zone41_Swansea_AQActionplan_1.pdf> [accessed 10 February 2022]

Dunnett, Nigel, and James Hitchmough. 2004. The Dynamic Landscape: Design, Ecology and Management of Naturalistic Urban Planting (London: Spon Press)

Gandy, Matthew. 2020. 'Urban Nature', in The SAGE Handbook of Historical Geography, 2 vols (London: SAGE), II, pp. 620–34

Gilbert, O.L., and Penny Anderson. 1998. Habitat Creation and Repair (Oxford: Oxford University Press)

Goode, David. 2014. Nature in Towns and Cities, New Naturalist Library (London: Harper Collins)

Goskar, Tehmina. 2012. 'Cu @ Swansea: An Industrious Future from an Industrial Past' <https://www.academia.edu/1969069/Cu_at_Swansea_An_industrious_future_from_an_industrial_past> [accessed 27 November 2021]

Griffiths, Ralph A. (ed.). 1990. The City of Swansea; Challenges & Change (Stroud: Alan Sutton)

Hilton, K. J. (ed.). 1967. The Lower Swansea Valley Project (London: Longmans)

Holt, G., H.E. Evans, and R.E. Davies. 1966. Tips and Tip Working in the Lower Swansea Valley, Lower Swansea Valley Project Study Reports (Swansea), West Glamorgan Archive Service, LOC SWA/222

Hughes, Stephen. 2008. Copperopolis: Landscapes of the Early Modern Industrial Period in Swansea (Aberystwyth: Royal Commission on the Ancient and Historical Monuments of Wales.)

Jarrige, Francois, and Thomas Le Roux. 2020. The Contamination of the Earth: A History of Pollutions in the Industrial Age, History for a Sustainable Future (Cambridge, Massachusetts: MIT Press)

Lachmund, Jens. 2017. 'The City as Ecosystem: Paul Duvigneaud and the Ecological Study of Brussels', in Spatializing the History of Ecology: Sites, Journeys, Mappings (London: Routledge), pp. 141–61

Lavender, Stephen J. 1981. New Land For Old: The Environmental Renaissance of the Lower Swansea Valley (Bristol: Adam Hilger Ltd)

Mather, A. 1991. 'Pressures on British Forestry Policy: Prologue to the Post-Industrial Forest?', Area, 23.4: 245–53

Owens, Nerys Elisa. 2009. 'The Shifting Governance of State Forestry in Britain: A Critical Investigation of the Transition from Productivism to Post-Productivism' (Cardiff: Cardiff University)

Pugsley, D.J. 1941. 'A Study of the Colonisation and Subsequent Flora of Coal Dumps at Cwmbach, Aberdare', The Proceedings of the Swansea Scientific and Field Naturalists' Society, II Parts 4 & 5: 159–77

Rackham, Oliver. 2015. Woodlands (London: Harper Collins)

Rees, Ronald. 2000. King Copper: South Wales and the Copper Trade 1584-1895 (Cardiff:

Reynolds, Ben, and Gordon Gibson. 2015. Swansea Eastside Connections, p. 60

South (Wales) Conservancy. 1952. History of Margam Forest 1921-1951 (H.M. Forestry Commission), Forestry Commission Archive

Teagle, W. G. 1977. The Endless Village: The Wildlife of Birmingham, Dudley, Sandwell, Walsall and Wolverhampton (Shrewsbury: Nature Conservancy Council)

Webb, John S., Iain Thornton, Michael Thompson, R.J. Howarth, and P.L. Lowenstein. 1978. The Wolfson Geochemical Atlas of England and Wales (Oxford: Oxford University Press)

Wilmot, Alzena, and Katy Harris. 2009. Community Woodland Baseline Report Wales (Edinburgh: Forest Research)

Right: The steam engine spring still survives above White Rock. Uncharitably described as 'Pond No. 4' in the ecology reports, the spring was the main source of water for the hill above White Rock and was connected to the steam engine site by a track and a ford on Nant Llwynheiernin. Now in a new guise as a biodiversity hotspot on the hillside (2022). (Author's collection).

Six: Nature and Restoration

For the moment, the land that formed the White Rock tips is highly regarded by a few people as either part of the industrial heritage park (what was the Cae Morfa Carw meadow) or the Kilvey woodlands. Although the tip over the riverside Cae Morfa Carw is currently only landscaped and grassed, the White Rock tip has been transformed into vibrant woodland. A generation of outdoor enthusiasts is growing up using Kilvey woodland with little or no idea of the troubled past of the land. Incredibly, walking across the hill, it is now hard to comprehend the area's industrial history.

As mentioned earlier, the considerable efforts to remove the tips are part of the story of the Lower Swansea Valley Project. Although the main tip was geographically outside the project's main aim of repairing the valley, removing what was then seen as one of the most visible tips above the valley was considered an essential addition. By 1968, after about six months of work, tip material was either removed or regraded and the lower part of the Nant Llwynheiernin valley, which had been covered under thousands of tonnes of slag, was reinstated. The Forestry Commission completed further ploughing and land treatment in the early 1970s to plant a crop of timber trees (Broad 1979: 10, 25–26). This additional ploughing was generally assumed to have disrupted much of the land surface, including removing the remains of the original pre-industrial roads across the land. Several medieval roads and a few substantial springs that supplied water for residents (and later the tip workers and their steam engine) certainly disappeared. By 1975, nobody believed any original land surface or features survived, although that is not the case. Walking the hill to assess woodland health, I could see that most of the original eighteenth-century leat system persists, a fact confirmed by the Lidar record.[1] Despite the evidence of considerable historical stream and leat networks, a much smaller volume of water runs off the western side of Kilvey. Nant Llwynheiernin, despite the size of its river valley, and its past importance for water power, barely registers a trickle in winter and is mainly dry the rest of the year. The same is true for the numerous leats adjoining the Nant.

Groundwater levels have changed dramatically since the early 1700s leaving many dry riverbeds, probably a result of coal mining draining water away on the north and east sides of Kilvey. The extensive leat network created between the

Left: Probably the biggest and oldest tree on the hill. A Monterey Pine planted in 1970 by Swansea County Borough Parks Department. Now increasingly rare in their native home of northern California, the Monterey has a special place in South Wales environmental history. It is resilient in a coastal location, resistant to fire, drought tolerant and thrives on industrial waste. This small stand of pines was planted on land contaminated by slag and pollution run-off above White Rock. It deserves its status as a Locally Significant Tree within the Kilvey woodlands. (Author's collection).

1730s and the 1820s may have been to improve quantities of water running into White Rock in the face of declining water flow from the natural stream network. The leats expanding further south to collect more streams that would have flowed down into Foxhole. This also explains the critical clauses in the copper works leases securing access to water resources in both leases discussed earlier. At least five significant channels above Foxhole connect streams or intercept watercourses and channel their flow into Nant Llwynheiernin.

The Lower Swansea Valley Project hugely benefited western Kilvey, and the restoration of woodlands in the 1970s and the later natural regeneration of vegetation are creating an environmental renaissance for Kilvey. We will never recreate the original forest on the Hill; we lost that centuries ago. Still, a combination of sensitive and thoughtful management from Swansea Council, NRW, and the local volunteer groups created a new type of urban Welsh woodland. Although Kilvey's future is still uncertain in the face of a new set of Council-led commercial changes that will irretrievably alter the woodland character of the hill.

So, the woodlands of Kilvey Hill are barely forty years old. They came from nothing, or more accurately, they were created out of the heavily polluted wastelands created by the White Rock copper works and the other primary Lower Swansea Valley polluters.[2] Kilvey dominates Swansea, and most residents know of it because it forms the backdrop to the city, it is intimately related to the history of Swansea. However, as it's on the east side of the River Tawe, its official inclusion within the administrative boundaries is as recent as the nineteenth century (Griffiths 1990: 53). Today, the Hill is primarily green and covered with trees both coniferous and deciduous. The urban curse of frequent fires punctuates the dry months, sadly typical for much Welsh woodland and a strangely Welsh phenomenon if recent researchers are to be believed (Llewellyn and others 2019: 805). For Swansea natives, Kilvey Hill dominates the city's central space, albeit alongside its more urbanised western neighbour of Townhill.

Kilvey undoubtedly started as a heavily wooded coastal hill overlooking the bay and river mouth. It most likely retained that character in post-Roman times and up to the years of Viking settlement at the beginning of the eleventh century (Linnard 1982: 10–19). The woods were gradually removed after 1100 supplying building timber and smaller woodland artifacts for local agriculture and the mining industry which was flourishing by the 1300s. The hill had many vigorous streams, so it is no surprise that the Knaploth corn mill originally above the White Rock site was an early arrival. The bigger trees of the primary woodland may have disappeared by the time the Normans arrived in the 1100s. The secondary woodland of smaller trees would have gone through successive cycles of growth and destruction in response to the fortunes of early Swansea in a cyclic process of growth and harvesting that is now thoroughly understood (Kowarik 2005: 4–7). The removal of larger trees tended to release a mass of new undergrowth, although it never reached the size and scale of ancient woodland as it was harvested quickly for

Right: One of the original 'Gasworks Pines' planted in 1970, still flourishing amongst later coniferous plantation that isn't doing so well in early 2022. (Author's collection).

local economic activity (Rackham 2022: 38–54).

The early exploitation of the Kilvey coal deposits meant the Hill is honeycombed with over 700 years of coal workings. Earlier ones are shallow scrapes into the sides of Foxhole and above White Rock and are still there now as secluded woodland glades. The sophistication of the mining industry demanded more productive coal pits in the early 1700s and the scars along the western side of Nant Llwynheiernin and the coal tips above the source of the Nant show increased exploitation of the important Hughes and Foxhole coal veins. The prevailing westerly winds guaranteed that Kilvey took most of the pollution from the valley, destroying the ecosystem and ruining the land. Since 1970, Kilvey's environmental condition has gradually recovered to what we see today.

Kilvey Hill is important to the environmental history of Swansea because of the dreadful way Lower Swansea Valley industries exploited and ecologically destroyed the land. The industrialisation of the past three centuries will continue to have implications and shape local geography for the present and future. Kilvey was an example of how bad pollution could be, how cynical and uncaring local industrialists could be, and, from a 1790s viewpoint, how far-reaching industrial pollution can be into the futures of unborn generations. Equally, Kilvey has become a focus of the beneficial outcomes of repair and restoration, albeit picked up by the public sector when private industry in the form of the Vivians and other industrialists walked away from their responsibilities in the depression years of the twentieth century (Griffiths 1988: 60–62). The greening of the Hill in the past two decades has been remarkable, particularly as it is based on the toxic lifeless wasteland of the 1940s.

In her explorations of space and time, Doreen Massey talks of spaces and their constant reconstruction as they undergo various trajectories of use and engagement, creating different types of history and geography dependent on politics, economics and changing social pressures (Massey 2005: 4–7). The decolonisation movement in Wales is a good example of critical re-evaluation of historical events, and Swansea's copper industry is an obvious place to start as so much Swansea copper was used to sheath the hulls of British slave ships or create commodities for sale in slave markets (Solar and Rönnbäck 2015; Museums Association 2021). Massey's trajectories were shaped by contemporary geographical issues, but are very useful in examining historical issues, as Massey herself discussed (Massey 2005: 62–63). Applying trajectories to the Kilvey story gives us a series of story threads, some conventional, others less so:

- Waterpower and the early corn mill.
- Mansel's coal business at White Rock wharf.
- The earliest coal mining on the Hill.
- The money and the Bristol slave trade.
- The siting of the copper works.
- The early waste problem.
- The early copper smoke problem.
- Waterpower engineering across the Hill.
- The arrival of steam engines.
- The Victorian pollution horror.

Above: Unmapped coal tips cover Kilvey. Most tips were abandoned before mapping began in the early 1800s. Weathering has meant that the tips are plant-friendly but their unconsolidated nature means that trees are easily uprooted once they reach a significant size. The coniferous plantations are particularly susceptible because of poor root-plate development. (Author's collection).

- The domination of the industrialists.
- The abandonment of responsibility for pollution.
- The acceptance of responsibility for the clean-up.
- The beginnings of repair and restoration.
- Recovery and the 'presencing' of nature.

Some of these are easy to follow in existing literature and the best place to start are the two classic volumes of Swansea history produced in the 1990s (Griffiths 1990; Williams 1990). Two other books are also very helpful. The first is a comprehensive, although technical, catalogue of the industrial landscape of Swansea produced by the Royal Commission on the Ancient and Historical Monuments of Wales (Hughes 2008). Secondly, there is an interesting re-interpretation of the town's history in its early industrial years, conventionally seen as the late eighteenth to the mid-nineteenth centuries (Miskell 2006). Between them, these four books give a wonderful view of the scope of Swansea's history and often, how we got to where we are now. Other stories are less easy to follow, for example, the pollution story and the treatment of the environment are rarely covered. A geographer must work harder to find knowledge in these topics. It is, perhaps, one of philosopher Bruno Latour's less controversial statements when he said 'History is no longer simply the history of people, it becomes the history of natural things as well' (Latour 1993: 82), a task often picked up in historical geography (Heffernan and Morin 2020: 25–40).

In trying to understand the development of green and ecological politics, one of the most significant books of recent times looks at the urban ecology of Berlin. It's not an unusual stretch to look at Berlin. We know a lot about urban ecology across Europe because of the pioneering work of botanist Herbert Sukopp and his researchers (often trained amateurs) in investigating nature in post-war Berlin. In the early 1960s, Sukopp started his programme of ecological site surveying that identified the significance of urban nature around the bombsites, parks and nature reserves of Berlin. Sukopp's research formed the basis of much early knowledge of habitat and restoration repair, being regularly quoted and referenced, for example in O. L. Gilbert's landmark work on British urban habitats and later work understanding types of urban vegetation (Gilbert 1989; Kendle and Forbes 1997: 29–30). At the same time, broadly similar work was being undertaken as part of the initial investigations for the Lower Swansea Valley Project. Berlin became a European focus for understanding urban nature, pollution, and ecology...but in many ways it could have been Swansea.

Taking inspiration from the mass of ecological knowledge collected in Berlin by Sukopp, the sociologist Jens Lachmund sought to understand the realities of the changing place of nature in a modern city (Lachmund 2013). In cataloguing over thirty years of political struggle in raising the prominence of nature and conservation, Lachmund charted the changes in social mood and local politics that allowed nature conservation to emerge as a prominent issue. He saw the transition from the conservation ecology of the 1960s to the urban nature conservation of today (Lachmund 2013: 224–26). He clearly saw something that we experience on Kilvey today, namely that the natures that interested traditional conservationists were not necessarily

Above: Kilvey Woodlands in the winter of 2021. The toxic slag dumps are shown in brown and purple in the north west corner of the land. The green land here is a culmination of land acquisition and management by Forestry Commission and Natural Resources Wales between 1970 and 1990. The habitats were a mix of coniferous woodland and grassland. In the summer of 2022, the leasehold of the woodland was surrendered back to Swansea Council and the future of the woodlands is now uncertain. (QGIS compilation).

the same issues in urban nature which enthused the local residents, who are in close proximity to the day-to-day environment (Lachmund 2013: 226–27). In Berlin, areas of land casually regarded as 'wasteland' developed remarkable and often luxurious stands of vegetation, but equally, these places were also recognised and valued as informal recreation spaces. Eventually, after considerable and bitter political conflict, local authorities were compelled to recognise both these features and incorporate them into local planning. The poor relationships between Swansea citizens and their local authority suggest here there is less confidence in that happening.[3] In Berlin, the conservation versus recreation debate became a significant source of tension between local communities. (Lachmund 2013: 225–26). A similar situation is evolving on Kilvey as tension emerges on the local authority plans for Kilvey's development and the conflicting opinions of the growing group of recreational users who are increasingly appreciative of the opportunities within the green space ('Lower Swansea Valley Forest Resource Plan - Natural Resources Wales Citizen Space - Citizen Space' [n.d.]). Lachmund's detailed conclusions are fascinating, finding that there was invariably conflict between land users who saw land in terms of amenities, and ecologists who only saw species and conservation. Eventually ecologists were often regarded as 'unreliable friends' in terms of forming common goals of conservation (Lachmund 2013: 230). The leaked news in the autumn of 2022 that Natural Resources Wales has surrendered its lease of the Kilvey Woodlands added new concerns to the wellbeing of Kilvey as it lost its protective status of being an important part of Welsh national forestry and moves towards being owned and occupied by a foreign leisure company.

Geographer Steve Hinchliffe explored the presencing (or re-presencing) of nature in the modern world (Hinchliffe 2007). Hinchliffe sought to acknowledge the multiplicity of 'natures' on a site. Helpfully he uses a woodland as one of his examples. He also references Massey's concepts of 'time-spaces' and trajectories mentioned earlier. Much of Hinchliffe's work is in response to some of Bruno Latour's dramatic views on nature where he looks to abolish the concept of nature because nothing in the world is now untouched by human influences. It is a powerful argument when considered in detail in his book which is still a highly sought after bestseller (Latour 2004). Happily, Hinchliffe presents a far more optimistic viewpoint of nature than Latour would allow us, however the recovering nature of Kilvey is totally built on the actions of earlier generations, a fact I constantly see re-confirmed whenever I see a blown over tree and look at the industrial waste it was growing on in the hole it leaves behind, a reality that is often forgotten by a new generation of politicians and planners, giving new life to the shifting baseline syndrome prevalent in so many environmental issues, as people develop a lack of awareness of the issues faced on Kilvey in the past (Soga and Gaston 2018).

Geographers often talk about the concept of

Right: Oak seedlings appearing on coal tip debris above White Rock in 2022. The reappearance of native species has been slow and perhaps a phenomenon only gathering pace over the past five years. There are no signs of older or larger native trees on the hill yet. It is possible that trees are still vulnerable to the pollution that lies in a layer in the subsoil. Nevertheless, the new natural regeneration that is appearing is a sign of progress. (Author's collection).

urban nature (Gandy 2020: 620–34). The idea of relationships between people and land can be a central theme, particularly in historical geography, where we can see the role of a space like Kilvey change over a thousand years from a centre of natural resources, to space for tipping industrial waste, a place of industrial devastation, a centre of recreation, and recently a focus for ecosystem services across much of the east side of the city. Geographers have found that Ecosystem Services can be notoriously challenging to assess, but guidance and evaluation improve as experience in data gathering improves (Everard 2017: 112–37). Recent UK parliamentary investigations also recognise the growing need to acquire more quality data to help understand the role of biodiversity in urban nature (Wentworth 2022). Even a small area like Kilvey is challenging to understand with complex relationships between nature, people and patterns of use. The problems are not new, foresters went on record to express their nervousness about establishing a forest on the hill near urban areas at the time of the establishment of the Kilvey woods in 1970, requiring a new set of understandings of the elements of urban woodland (Broad 1979: 10–11). Nowadays, there is an amount of justifiable pride in the now-defunct Forestry Commission's work establishing the unique urban forests in Wales on Kilvey Hill and the adjoining land (Broad 1979: 11)

Between 2017 and 2022, a series of ecological surveys were conducted across Kilvey. In preparation for redeveloping Kilvey as part of an adventure tourism project. The surveys were conventional ecological surveys conducted as part of a preliminary planning application process. Although necessarily limited in extent, they were thorough, and covered the presence of birds, animals, some reptiles, and vegetation types. The reports catalogue incredible recovery (Kilvey Hill, Swansea: Ecological Appraisal 2023). Remarkably, in over a hundred pages of data there isn't a reference to pollution or the environmental problems of the hill. Remarkable because one interpretation of that fact is how the natural recovery of the land has pushed pollution into the background. It is a tribute to the generations of council workers, forestry staff and volunteers who have worked to transform Kilvey since the 1970s, and now we have evidence of the success of that effort.

If there is a problem with such ecological reports, it is in how they are perceived. Written for planning apparatchiks and specialists for a specific tick box exercise in planning permission, it can be a challenge to use them constructively. How do they help in understanding 'nature' on Kilvey? Not a straightforward question. Ecologists are compelled to understand habitat by using various classification schemes (basically simplifications). Much of Kilvey gets lumped together as 'Coniferous Woodland' or 'Semi-natural Woodland', Conifers in particular are usually considered undesirable and now mostly shunned in Welsh Government policy in favour of 'Native Woodland' (Welsh Government 2018: 44–53). Classifications of habitats are useful, and they can provide a broad picture of the sort of plants and animals to expect, although they can never provide true detail (Averis 2013: 378–83). So, do the reports help us understand the nature of Kilvey? As discussed earlier, much of Kilvey's coniferous woodland was planted because little else would grow, and the 1970s Forestry Commission only thought of forests as collections of commercial species. It's a bit like the old saying, 'if the only tool you have is a hammer,

Above: Although only one of many condition indicators in woodland, the presence of birds is always of interest. This is an extract from a remarkable series of records kept by volunteer warden Karl Squires between 1997 and 2004. Numbers of birds on a site are often unreliable but the species count appears to have risen slowly. In the late 1990s, Karl recorded forty three species of bird across the top of Kilvey and into the woodland. The more recent survey of 2019-20 saw an increase of species to fifty one.

Analysis of presence/absence is challenging in urban woodland but there are strong presences of birds that prefer dense low shrubs (often fire damaged areas) such as Willow Warbler, Dunnock, Marsh Tits and Bullfinches. The lack of suitable veteran tree habitat has made it difficult for owls to get much of a hold on the hill. The relatively open canopy of some of the northern woodlands above White Rock has encouraged Spotted and Pied Flycatchers, Redstarts, and some woodpeckers. Nightjars and Goshawks have been common from Karl's earliest surveys to the present day. Perhaps the most optimistic view is that although purists tend to disregard the biodiversity value of the coniferous woodland, the resident bird species are less discriminatory and species diversity is increasing. Wild animals and birds rarely read the ecology handbooks and tend not to follow rules, particularly in recovering urban woodlands. (Data from Karl Squires' personal records 2022).

everything looks like a nail'. Ironically, the ability of thick coniferous woods to create quiet seclusion rapidly is one of the highly valued features of the 2023 Kilvey Woodland. The poisons of the land are now taking a toll on tree health, exacerbated by poor forest management over the past twenty years. A new woodland management regime fit for the future is sorely needed.

One of the priority habitats that the Welsh Government want to protect is 'Ancient Woodland' (Palma 2014). As always, there has to be a classification, and woodland in Wales is regarded as 'ancient' if it has been around for about four hundred years (Welsh Government 2018: 7). There are also many woodland plants that sometimes indicate the presence of old woodland such as buttercups, bluebells, wild garlic and many more (Field Studies Council 2016). Still, identification of ancient woodland across industrial Glamorgan is challenging and elsewhere, I have seen woodlands that still show the devastation of coal mining from the 1700s classified as 'ancient' on the tenuous presence of a few plant species. There is a general expectation that planting a grove of trees equates to recreating something from the past. That can never be true in recreating woodland. We cannot recreate the ecological community around a thousand-year oak tree by planting a stand of new seedlings. That is the worst kind of planning greenwash. However, we can have an approximation of ancient woodland and hope for the best. The noted scholar Oliver Rackham found several ancient woodlands across eastern Glamorgan that he estimated to have survived destruction in the sixteenth and seventeenth centuries, tracing the complexity of tree species and woodland plants in remote areas. However, he never investigated Swansea as he probably realised how industrial it was and assumed there was nothing here (Rackham 2022).

Kilvey's trees were the first casualties of copper smoke pollution in the 1700s with native trees being particularly susceptible to the poison in the air. It is highly likely that the Scots Pines planted in 1970 above White Rock were the first to be planted (and survive) for nearly three centuries. Between the 1750s and 1970, there were no trees, so for a long time the essential components of a functioning healthy woodland were nowhere on the hill. One of the biggest absences is old trees and the dead wood they provide. Old trees have scars and holes and shed branches supplying shelter and food for woodland animals. Kilvey will take another century to generate older trees that, in turn supply shelter for woodland species. It's a slow process and it's only now getting started as conifers die and are replaced naturally by native woodland. The toxic soils make the process even slower.

So, Kilvey is not Ancient Woodland, a disappointing reality to some. However, what is there is the coniferous woodland planted in the 1970s, with recovering natural mixed woodland that has grown in recent years in the clearings of fire damage and blown-over conifers. A mixture of conifers, non-natives such as Rhododendron and Japanese Knotweed, native ash and oak, and a host of underwood plant species described as 'spontaneous' vegetation (Gilbert 1989: 302–10). Again, it's worth looking at the Berlin ecologists to understand the concept of 'industrial nature' where vegetation appeared 'spontaneously' on bombsites and railway yards after 1945. The vegetation became of interest because of its diversity and resilience. More study of the plant groupings and communities led to a broader concept of urban nature across many European industrial sites (Storm 2014: 5). But, we don't have to go as far afield as Berlin. After World War Two, Swansea's bombsites started producing their own exotic vegetation communities. By 1946, Swansea botanists recognised something unusual was happening where all sorts of grassland and woodland plant species were coming together in new 'bombsite' communities, and recorded the incredible diversity of plants growing across the urban centre of Swansea (Sykes 1947). The study of 'bombsite flora' also became a national if short-lived pursuit across the bombsites of London, Birmingham and Manchester (Bevan 2008). More recently, the ecologist Ingo Kowarik classified vegetation types into four broad categories, the first kind was plant remnants of the post-glacial landscape (we don't have any here), the second kind was plant remains from agricultural landscapes, the third kind were parks and gardens, and the fourth kind being 'spontaneous' vegetation adapted to polluted urban or industrial environments (Kilvey included) (Kowarik 2011). There will be non-native species in the mix, some of them are invasive and tenacious. Japanese Knotweed is the traditional Glamorgan 'scary' plant, and there are several more. Kilvey is now becoming infected with Rhododendrons, although these are also widely sold as decorative plants, so it is hard to police such things. Regardless of their status, these non-native plants are here to stay and are now part of the mix of local plants you see in urban woodlands (Cooley 2022). The inference from these various studies being that regardless of broader classifications, local environmental conditions of weather, pollution, soil type and land use create a locally appropriate natural grouping of plants. On Kilvey we now see 'spontaneous' oak trees and other plants emerging in clearings where conifers have fallen or burning has ocurred. However, the absence of larger oak trees suggests this is either a new phenomenon or the trees are still susceptible to copper pollutants in the soil and die after a few years. We don't yet know.

If Kilvey is not ancient woodland it is something else. We know the coniferous woodlands were planted in the 1970s. Everything that is there now has happened since then, so there is no continuity with the past. Whilst this is not nationally unique, it is at least distinctively Welsh, and forms a common thread with the coal tip devastation of Welsh valleys in the nineteenth-century. The phenomena of post-industrial spontaneous vegetation was researched far earlier in Wales than the bombsites of Berlin, David Pugsley conducted pioneering work in the 1930s. Pugsley looked at coal tips near Aberdare that had been set down in the 1880s, leaving us with foundation knowledge that has helped us understand much about how plant life responds to pollution (Pugsley 1941). On Kilvey, trees, plants, insects, and animals all started to arrive as food and shelter allowed. What we can experience on the hill is the response of local ecology to the evolving environmental conditions. Local wildlife, often threatened, homeless and constantly attacked by people, cars, and pollution moved into the space on the hill from wherever it was struggling to live. The plant and animal community within Kilvey Woodland is an assembly of local wildlife that had survived the years of industrial devastation. Kilvey is a living laboratory for urban and industrial recovery and wider climate change. It is a distinctive post-industrial ecological community living in a distinctive Welsh post-industrial woodland. In the rush to redevelop the land there is a dreadful risk that the uniqueness of the land, still only forty years old, will be lost in the destruction, revision and replanting by developers intent on creating an alien vision based on characterisation of 'what should be there' with alien landscape architectural values, commercial requirements, and unhelpful ecological generalisations, turning a uniquely Swansea landscape to 'anywhere in Wales'.

Notes

1. ('Historic LIDAR Archive' 2021).

2. Particularly, White Rock, Morfa, Hafod, Landore, Fforest, Middle Bank, Upper Bank, Ynys and Rose copper works 1737-1928.

3. In conversations with local politicians and officials in the autumn of 2022, I was regularly told that Kilvey was a 'problem area' that needed a solution. A disappointing viewpoint considering the investment work of the past forty years of ecological restoration.

REFERENCES

Averis, Ben. 2013. Plants and Habitats: An Introduction to Common Plants and Their Habitats in Britain and Ireland (Edinburgh: Ben Averis)

Bevan, David. 2008. 'Out of the Ashes: The Greening of the City's Bombed Sites', in London's Changing Natural History: Classic Papers from 150 Years of the London Natural History Society (London: London Natural History Society)

Broad, K.F. 1979. Tree Planting on Man-Made Sites in Wales (Forestry Commission) <https://www.forestresearch.gov.uk/documents/6846/FCOP003.pdf> [accessed 23 September 2021]

Cooley, Hazel. 2022. Invasive Non-Native Species (Parliamentary Office of Science and Technology), p. 8

Everard, Mark. 2017. Ecosystem Services: Key Issues (Abingdon: Routledge)

Field Studies Council (ed.). 2016. 'Guide to Ancient Woodland Indicator Plants' (FSC Publications)

Gandy, Matthew. 2020. 'Urban Nature', in The SAGE Handbook of Historical Geography, 2 vols (London: SAGE), II, pp. 620–34

Gilbert, O.L. 1989. The Ecology of Urban Habitats (London: Chapman and Hall)

Griffiths, Ralph A. 1988. Singleton Abbey and the Vivians of Swansea (Llandysul: Gomer)

——— (ed.). 1990. The City of Swansea; Challenges & Change (Stroud: Alan Sutton)

Heffernan, Michael, and Karen M. Morin. 2020. 'Between History and Geography', in The Sage Handbook of Historical Geography, 2 vols (London: SAGE), pp. 25–40

Hinchliffe, Steve. 2007. Geographies of Nature (London: SAGE)

'Historic LIDAR Archive'. 2021. <https://libcat.naturalresources.wales/folio/?oid=116826> [accessed 1 December 2021]

Hughes, Stephen. 2008. Copperopolis: Landscapes of the Early Modern Industrial Period in Swansea (Aberystwyth: Royal Commission on the Ancient and Historical Monuments of Wales.)

Kendle, Tony, and Stephen Forbes. 1997. Urban Nature Conservation: Landscape Management in the Urban Countryside (London: E & FN Spon)

Kilvey Hill, Swansea: Ecological Appraisal. 2023. (Cardiff)

Kowarik, Ingo. 2005. 'Wild Urban Woodlands. Towards a Conceptual Framework', in Wild Urban Woodlands (Berlin Heidelberg: Springer-Verlag), pp. 1–32

———. 2011. 'Novel Urban Ecosystems, Biodiversity, and Conservation', Environmental Pollution (Barking, Essex: 1987), 159.8–9: 1974–83 <https://doi.org/10.1016/j.envpol.2011.02.022>

Lachmund, Jens. 2013. Greening Berlin: The Co-Production of Science, Politics, and Urban Nature (Cambridge, Massachusetts: MIT Press)

Latour, Bruno. 1993. We Have Never Been Modern (Harvard: Harvard University Press)

———. 2004. Politics of Nature: How to Bring the Sciences into Democracy (Harvard: Harvard University Press)

Linnard, William. 1982. Welsh Woods and Forests: History and Utilisation (Cardiff: National Museum of Wales)

Llewellyn, David H., Melanie Rohse, Jemma Bere, Karen Lewis, and Hamish Fyfe. 2019. 'Transforming Landscapes and Identities in the South Wales Valleys', Landscape Research, 44.7: 804–21 <https://doi.org/10.1080/01426397.2017.1336208>

'Lower Swansea Valley Forest Resource Plan - Natural Resources Wales Citizen Space - Citizen Space'. [n.d.]. <https://ymgynghori.cyfoethnaturiol.cymru/forest-planning-cynllunio-coedwig/lower-swansea-valley-forest-resource-plan/> [accessed 16 November 2022]

Massey, Doreen. 2005. For Space, 1st edn (London: SAGE)

Miskell, Louise. 2006. 'Intelligent Town': An Urban History of Swansea, 1780-1855, Studies in Welsh History (Cardiff: University of Wales Press)

Museums Association (ed.). 2021. 'Supporting Decolonisation in Museums' (Museums Association)

Palma, Adriana De. 2014. Ancient Woodland (London: Houses of Parliament) <https://researchbriefings.files.parliament.uk/documents/POST-PN-465/POST-PN-465.pdf>

Pugsley, D.J. 1941. 'A Study of the Colonisation and Subsequent Flora of Coal Dumps at Cwmbach, Aberdare', The Proceedings of the Swansea Scientific and Field Naturalists' Society, II Parts 4 & 5: 159–77

Rackham, Oliver. 2022. The Ancient Woods of South-East Wales, ed. by Paula Keen (Dorset: Little Toller Books)

Soga, Masashi, and Kevin J Gaston. 2018. 'Shifting Baseline Syndrome: Causes, Consequences, and Implications', Frontiers in Ecology and the Environment, 16.4: 222–30 <https://doi.org/10.1002/fee.1794>

Solar, Peter M., and Klas Rönnbäck. 2015. 'Copper Sheathing and the British Slave Trade', The Economic History Review, 68.3 ([Economic History Society, Wiley]): 806–29

Storm, Anna. 2014. Post-Industrial Landscape Scars (New York: Palgrave Macmillan)

Sykes, M. H. 1947. 'The Flora of the Bombed Areas and Slum-Clearance Sites of Swansea', The Proceedings of the Swansea Scientific and Field Naturalists' Society, II.8 & 9: 291–306

Welsh Government. 2018. Woodlands for Wales Strategy (Cardiff: Welsh Government)

Wentworth, Jonathan. 2022. 'Biodiversity Indicators' <https://post.parliament.uk/research-briefings/post-pb-0041/> [accessed 10 February 2022]

Williams, Glanmor (ed.). 1990. Swansea: An Illustrated History (Swansea: Christopher Davies)

Above: The parcels clause of the 1736 lease which is transcribed on the following pages. The description of the land at White Rock follows the standard phrase 'All that' on the top line. The draughtsman clearly had some difficulty understanding some of the Welsh place names. The reference here to 'White Rock' refers to the coal wharf that was south of the current site of the White Rock copper works.

Annex 1: The original White Rock lease of 1737

This is a transcript of the original copy lease abstract currently held in the West Glamorgan Archive Service. This is the 'office copy' which was held by the landlord's lawyer, which explains why it has survived. The office copy was a direct copy of the original lease (described as an 'Indenture'. As always with a lease, there is a mass of information that has to cover every foreseen eventuality that the lawyers could think of, plus some cover for circumstances they could not have foreseen. Title deeds are vital for landscape history although, as always, there is the health warning that the survival of a handfull of documents only gives us a narrow insight into what happened in that small bend of the river. I have added explanatory notes throughout.

ABSTRACT of Lease
from the Honorable Bussy
Mansel Esq.[1] to Thomas Coster
Esq.[2] and others of Copper Works in the
County of Glamorgan for 51 years from
Lady Day 1737
ABSTRACT of Lease from the Honorable
Bussy Mansel
Esq. to Thomas Coster Esq. and others of
Copper Works in the County of Glamorgan for 51 years
From Lady Day 1737

Indenture of this date between the Honorable Bussy Mansel of Britton Ferry
In the County of Glamorgan Esq. of the one part and Thomas Coster of the City of
Bristol Esq Joseph Percival and Samuel Percival of the same City Merchants
Henry Barne of the City of Bristol aforesaid of the other part
Reciting that by certain Articles of Agreement indented bearing date the
twenty fourth day of August then last past made between the said Bussy
Mansel of the one part and the said Thomas Coster Joseph Percival and
Samuel Percival of the other part the said Bussy Mansel for the considerations
therein mentioned Did for himself his heirs executors and administrators
covenant promise and agree with the said Thomas Coster Joseph Percival
and Samuel Percival their executors administrators and assigns that he
the said Bussy Mansel would on or before the 1st of November then next
ensuing at the Costs of him the said Bussy Mansel by good and effectual
Leave and Grant in the Law as Counsel should advise demise and grant
unto the said Thomas Coster Joseph Percival and Samuel Percival their exors
Admors and assigns All that parcel or piece of waste or rough ground
containing or to contain 6 acres more or less and partly known by the

1. Bussy Mansel, the fourth Baron Mansel.
2. Thomas Coster inherited from his father, a Bristol industrialist, the Upper Redbrook copper works at Bristol, together with copper and tin mining interests in Cornwall. Coster is the link to slavery and Atlantic trading.

name of Craig Knap Loth and White Rock Coal place or Banks [3] and all other
the Lands Mill privileges and advantages thereinafter particularly mentioned
and expressed for the several purposes and for the term thereinafter ment
And reciting that since the execution of the said therein recited Articles
the said Thomas Coster Joseph Percival and Samuel Percival had admitted
and taken the said Henry Barne to be a partner with them to be a partner with them in the works
Business and Undertakings thereinafter ment(ioned)
It is witnessed that the said Bussy Mansel in conson of the yearly rents
Covenants Articles Provisos and Agreements thereinafter mentioned and contained
on the parts and behalfs of the said Thomas Coster Joseph Percival Samuel
Percival and Henry Barne their executors admors and assigns to be paid

Page 2

performed fulfilled and kept and in pursuance and full performance of the
said recited Articles of Agreement and of all other Agreements whatsoever made
or entered into between the said Bussy Mansfield (sic) Thomas Coster Joseph Percival
and Samuel Percival or any of them relating to the Land Mills privileges
and Advantages thereafter mentioned Did demise grant and to farm let and set
unto the said Thomas Coster Joseph Percival Samuel Percival and Henry
Barne their exors admors and assigns

All that parcel or piece of waste or rough Ground [4]
containing or to contain Six acres more or less called
or known by the name of Craig Knaploth and White
Rock Coal place or Bank with free liberty to erect
and build the several Buildings thereon thereinafter
mentioned and also so much of that Close on
Morva Carw as the said Thomas
Coster Joseph Percival Samuel Percival and Henry
Barne their exors admons and assigns should
have occasion for or want for laying of Slaggs
Cinders or other Rubbish proceeding and to proceed
from the intended works and workhouses
thereinafter mentioned
And also all that old decayed or ruinous Water
Corn Grist Mill called Knaploth Mill [5] situate near
White Rock aforesaid with free liberty to pull down and
destroy the said Mill and to erect and build in the stead thereof any other Mill or Mills Engines or

3. The original drafting clerk struggled with Welsh placenames and anclicised or simplified whenever possible. He is referring here to Cnap Coch above the current White Rock. The coal banks used by the Mansels were slightly downstream of the current reconstructed dock at White Rock and carried the original name of the place which was adopted for the copper works themselves.
4. The beginning of the parcels clause that describes the property. It also mentions the field of Morfa Carw which became the field that was used to build Middle Bank copper works.
5. The original corn mill at the mouth on the Nant Llwynheiernin.

other [6]
devices whatsoever if wanted or necessary for the use
of the said intended Works thereafter mentioned
and all the free use and benefit of the Dock or
Key at White Rock aforesaid for landing and
laying down Oar (sic) Clay or other Goods and shipping
of Copper and other Metals or goods [7] And also all
Springs Streams Waters and Watercourses whats(oever) [8]
issuing out of or running over from or through the
Lands or pieces called Craig Tyle Brown Lanerch
Clyder Melyny Wrane or any other Lands of the
said B Mansel his heirs or assigns or the

Page 3
Lands of any other person or persons whatsoever
from whom the said Bussy Mansel his heirs
or assigns had should or might procure leave
licence or liberty for the same and that by any
ways or means whatsoever might be brought to
White Rock aforesaid with free liberty power and
authority for the said Thomas Coster Joseph
Percival Samuel Percival and H. Barne their
exors admons and assigns to be made and erect
any headwares Stanks Banks Ponds troughs
Gutters Cuts and other devices whatsoever for the
diverting carrying conveying supplying retaining
preserving and keeping of the Waters Springs and
Streams aforesd. to White Rock aforesaid and for
the use of the intended Works thereinafter
mentioned as they should think fit at their
wills and pleasures Together with all landing
places ways All which said premises
were situate upon or near the River Tawy
and elsewhere in the parish of Lansamlet
in the said County of Glamorgan and then
or late in the possession tenure or occupation
of the said Bussy Mansel his tenants or
Undertenants / Except and always reserved
out of the now abstracting Indenture unto
the said Bussy Mansel his heirs and assigns
a convenient and sufficient place and places
upon the said Bank called White Rock
near the said River Tawy for the laying

6. Engines here is all types of machinery.
7. Here is some indication of the need for the dock or shipping place. The ore could be from Cornwall, or Parys Mountain on Anglesey.
8. The first acknowledgement that water will be needed for power and other processes.

down and shipping of all such Coals and
Culm as he or they or any of them shod cause
work in the said psh of Lansamlet with free ingress
& regress for Carts Carriages to lay down unload [9]
& ship of the same And also the use of the sd
Dock and Key for such vessels to lie and moor in
as should take load the Coals Culm afsd not at any time
hindering Ships or Vessels to load and unload Oar (sic)
Copper or other Metals or goods for the use of
the said intended works at a Crane to be
made and fixed on the said Dock for that
purpose

To hold the same with the appurtenances together with the
liberties powers and authorities aforesd unto the said Thomas
Coster Joseph Percival Samuel Percival and Henry Barne
their exors admons and assigns from the 25th March next
ensuing the date thereof for the term of 51 years from thence
next and fully to be ended

Yielding and Paying unto the said B Mansell (sic) his heirs
assigns the rent or sum of £24 on the feast day of Saint Michael the
Archangel next ensuring the date thereof and the like sum of £24 on
the feast day of the Annunciation of the blessed Virgin Mary which
should be in the year 1738 and from and after the said feast day of
the Annunciation of the blessed Virgin Mary 1738 Yielding and Paying
unto the said B. Mansel his heirs or assigns for the remainder of the
term thereby granted the yearly rent or sum of £63 of lawful British
money on every the feast days of Saint Michael the Archangel and the
Annunciation of the blessed Virgin Mary by even and equal portions free
and clear of all Rates Taxes Tallages impositions and assessments by
Act of Parliament or otherwise howsoever which should happen to
fall due on the premises or any part thereof during the sd Term

And the said Thomas Coster Joseph Percival Samuel Percival
Henry Barne for themselves and their respective heirs exors admons and
assigns did covenant promise and agree to and with the sd Bussy
Mansel his heirs and assigns in manner and form following (that
is to say) That they the said Thomas Coster Jos. Percival Samuel [10]
Percival and Henry Barne their exors admons & assigns or some or one
of them should and would at their own proper costs and charges in all
things with all Convenient speed and within the space of 3 years then
next at the farthest in good substantial and workmanlike manner

9. The dock is expected to be busy and there will be conflicts between boats wanting space to load and unload between the tides.
10. Mansel wants the Costers to start building the works immediately to allow profits to begin.

with good and sufficient materials erect build and set up on the
said demised premises One new Copper Smelting house or workhouse
for Smelting making and refining Copper or any other Metals and to

Page 5

contain 20 furnaces at least with such Warehouses Mills and other Buildings
as they the said Thomas Coster Joseph Percival Saml. Percival and Henry
Barne their exors admors and assigns should judge necessary convenient
And should and would at their own proper costs and charges provide all
and all manner of timber (except the Timber thereinas covenanted and
mentioned to be given and delivered by the sd B. Mansel his heirs [11]
assigns for and towards carrying on the said Buildings Stones Bricks
Laths tiles Nails Iron Sand and Lime and all other materials which shod
be fit and necessary to be used in and about the said buildings and shod
[carry on the same with effect until the sd Buildings should] be completed and finished and should
and would therein (illegible)
expend and lay out the sum of £1400 of lawful British Money over
and above and exclusive of the value of the Timber to be employed
and used therein And in case the said sum of £1400 should not
be sufficient to finish the said intended works the said Thos. Coster
Jos. Percival Sam. Percival and Henry Barne their exors admons and assigns should
carry on and finish the same at their own proper costs charges
without any allowance from the said Bussy Mansel his heirs or ass(igns)
for or in respect thereof.
That they would during the said term of 51 years as often as need(ed)
should require at their own costs repair maintain uphold glaze and
keep the said houses and buildings to be erected and finished and
all other the premises thereby granted with all manner of needful and
necessary reparations whatsoever and at the expiration or sooner at
determination of the said term quietly leave and yield up together
with all furnaces Bars or binders to the same belonging and in
case any furnaces should be broken up or taken down which the
said Thos. Coster Jos. Percival Saml. Percival and Hy Barne their
exors admons and assigns had by the now abstracting Indenture a
right and power to do in order to take out the Bottoms [12] or on any
other account whatsoever they should at their own costs and charges
rebuild or sufficiently repair the same with new bottoms and leave the
same in sufficient state and condition for making smelting refining
of Copper or any other Metals by them usually made or melted unto

11. The Costers must procure their timber from the Mansel Estate. The Mansels had a lucrative timber trade and exported timber from White Rock to Bristol for clients such as the Royal Navy.
12. The bottoms are the hard clinker that settled at the bottom of the furnaces. It was generally very rich in copper and other metals. It was hard to process but potentially very profitable. I have seen iron industry bottoms (known as bears) across Shropshire fields where they were dumped as unworkable in the 1700s and are so hard and dense that they still survive. Copper bottoms would always be reworked because of their value.

the said B. Mansel his heirs or assigns at the end or other determin(ation)
of the said Term
That they would not at any time during the sd Term alien

Page 6
assign or dispose of the said premises to any person or persons who should
at that time be concerned or interested in any Copper work or Copper works
situate on the Rivers of Swansey or Neath without the licence of the sd
Bussy Mansel his heirs & assigns first obtained [13]
Covenant by the said Bussy Mansel that it should be lawful for
the said T Coster Jos. Percival Saml Percival and Harry Barne during the
sd term to dig raise take and carry away all such stones as they should
from time to time have occasion for towards the erecting and building such
Copper works houses and buildings and the furnaces therein to be made
on the said demised premises from and out of any Quarry and quarries of
Stones upon or under the Lands of the said Bussy Mansel his heirs and
assigns in Lansamlet aforesaid doing the least damage that might be to the Soil thereof.
That said Thos. Coster Joseph Percival Saml. Percival & Henry
Barne should at all times during the said term in any convenient places in
the Marshes or other lands of the sd Bussy Mansel his heirs or assigns
near to the sd premis(es) and where it might be least hurtful to the said Bussy
Mansel his heirs or assigns dig take and cast up so much Earth and Clay
out of the said Marshes and other lands as they should from time to
time make and convert into Bricks or Tiles and also such Earth Sand
and Loam as should be wanting for the use of the furnaces to be
built in the said intended work and also sufficient room and
place for burning such Bricks and Tiles and all Limestones and Lime
towards the building and repairing the said intended Houses
and buildings such Bricks Tiles Earth Loam and Lime to be used
only in and about the making building and erecting of new
Messuages houses mills furnaces or other conveniences upon the
premises or for repairing amending or bettering thereof or about
the smelting the said Copper and not for any other use
whatsoever without leave first obtained from the said B. Mansel
his heirs and assigns

That the said B. Mansel should furnish and deliver to Thos.
Coster Jos. Percival Saml Percival and Hy Barne all such Coals as
should be necessary for burning such bricks and lime at the rate of 18 [shillings]
per weigh to be delivered at the Brick or Lime kilns
That the sd T Coster J Percival S Percival Hy Barne shod as often as they shod have occasion to

Page 7
take and carry away from the lands of the said B. Mansel his

13. The Mansel's lawyer knew the economics of the metal industry and knew there was a risk that the Costers would sell the copper works onto other operators if they could. This clause stops that happening.

heirs and assigns in the parish of Lansamlet first giving notice to some [14]
known agent of the sd B. Mansel all such underwood (Oak Elm Ash and
Sycamore and hedges excepted as should be necessary for refining of Copper
at the sd intended Works without rendering or paying any thing
therefore so that the same did not exceed three long cords according to
the usual measure in any one year during the said term
That they should have free liberty during the sd term for laying down
the Slaggs cinders ashes and rubbish that should arise from the sd
works in any lands belonging to the sd. B. Mansel contiguous to the
sd intended works so that the same were not cast or known whereby to
prejudice the navigation of the River Tawy
That the sd B. Mansel in conson of the sd buildings to be erected
and towards carrying on the same and to make up the sd sum of
£1400 before covenanted to be expended in and about the same shod
pay unto the said T Coster J Percival S Percival and H Barne
their exors admons or assigns the sum of £800 on or before the 25th of that
instant March and should upon reasonable notice furnish and provide
such and so much good and merchantable wood and timber as shod
be necessary for erecting and building the sd intended houses and buildings
and carrying on and finishing thereof and deliver such wood and
timber at such convenient place or places near the same as should be
appointed by the agents or workmen of the said T Coster J Percival
S. Percival and H Barne employed thereabouts without expecting any thing
for such wood or timber so that the value of the same did not exceed
the sum of £250 after the rate of 35 [shillings] by the Ton for each ton of the
said timber so to be delivered as aforesd according to the usual measure
And in case the timber to be used and employed in the said buildings
should exceed the value of £250 after the rate of 35 [shillings] per ton Then that
the said T Coster J Percival S Percival and H Barne should pay
unto the said B. Mansel his heirs or assigns the sum of 35 [shillings] by the
ton for every ton of timber to be employed and used in the sd intended
buildings that should be over and above the said timber so as afsd
to be delivered to the value of the sd £250 and in case the said
T Coster J Percival S. Percval and H. Barne should not expend
and use in and about the said intended buildings so much timber

Page 8

as after the rate of 35 [shillings] by the ton should amount to £250 that then
the sd B. Mansel his heirs and assns. shod after the sd buildings should be

14. All industrial processes consumed large quantities of timber. This is one of the reasons why Glamorgan was largely denuded of trees by the end of the eighteenth century. Copper smelting needed large quantities of charcoal and this would be made as close to White Rock as possible as charcoal tended to collapse to dust if it was transported in carts. Small boats carrying charcoal across the river can be seen in a print of Forest Copper Works by Thomas Rothwell in 1791. What is surprising here is the acknowledgement that Sycamore is a timber tree of worth. The hedges of the Penllergare Estate were replanted with Sycamore in the early 1800s probably as a faster growing alternative to Oak.

finished pay unto the sd T Coster J Percival S Percival and H
Barne so much ready money as with the value of the timber that shod
be delivered for the purposes aforesaid after the rate of 35 [shillings] by the ton shod
make up the sd sum of £250
And also shod maintain & keep the Dock and Key at White Rock afsd
in good and sufft. repair dur(ing) the term
That he should dur(ing) the cont(inuance) of the sd term on reasonable notice furnish and
deliver at the sd intended workhouses for the use of the sd T Coster J Percival
S Percival and H Barne all such good clean and merchantable coals and [15]
such as were fit for copper furnaces as shod from time to time be requisite
reqd directly from such pit or pits as the foreman or refiner to be employed in
the said intended works should chuse [sic] and approve of without being picked
or culled of the Big Coal but as it shod be raised throughout at the rate of 21 [shillings] for
each weigh and to be measured at the sd intended works by barrows each barrow
to contain 50 Winchester Gallons and each weigh to contain 28 0f the sd
Barrows for all coals brought ot the said works (except the Coals brought)
from Popkins's or Knaploth Small Coal Vein [16] and in case the said
T Coster J Percival S Percival and H Barne shod at any time during
the said term be minded to have any quantity of coals from the said pit or
Veins called Popkins's or Knaploth Small Coal Vein the said B. Mansel
shod furnish them therewith at the said intended works clean and merchantable
as the same were raised at the rate of 18 [shillings] by the weigh for such coals each
weigh to contain the quantity and measure aforsd if the Coal works of the
said B. Mansel his heirs or assigns within the parish of Lansamlet aforsd
shod so long continue to be carried on and worked or might be wrought to profit
a sufficient quantity of coals wrought thereout whereby to supply the sd works
but in case the sd coal works of the sd B. Mansel shd not happen to be carried
on and wrought he or they should not be obliged to purchase coals elsewhere to
supply the sd works
That he shod upon reasonable request and notice furnish and deliver unto the
said T Coster J. Percival S Percival and H Barne all such timber and
other wood as they shod during the said term have occasion for making Lauder
Troughs (Except Troughs for conveying the water from Craig Tyle Brown to the sd
works / Dams floodgates mills wheels repairs and other necessary use in & about

Page 9

the premi(ses) at the rate of 35 [shillings] for every ton so to be delivered as afsd for the sev(eral) uses therein
last ment(ioned) and shod furnish & allow unto them when necessary dur(ing) the sd term all such

15. The quality of coal was of extreme importance and controversy. Coal merchants would always try and give cheaper coal, shale and small lumps to the industrialists and charge higher prices for the 'big coal'. Kilvey coal was typically bituminous or 'binding' coal but veins of poorer coal were also present. Poorer coal was sold for brewing or household use. The industrialists were very aware of this sharp practice.
16. The Knaploth vein was known to later generations as the Captain's vein. William Logan was still trying to understand the positions and qualities of the coal veins over a hundred years after this clause was written.

timber
as should be wanting for making troughs for carr(ying) the water from Craig Tyle Brown to
to the sd intended works at White Rock without expecting anything for the same
That he should upon request for that purpose supply them with such houses or
cottages for habitation & lands for their necessary uses near the said premises as they shod
have occasion for paying therefore as much yearly rent as the then present tenants or occupiers
thereof did pay for the same over and above the rent thereby reserved
That if it shod happen that the sd B. Mansel co(uld) not upon reasonable terms obtain a
Lease or Grant from his Grace the Duke of Beaufort of that parcel of nde Ground called [17]
Morva Carw adj(oining) to or near the sd premi(ses) That then the sd B. Mansel wo(uld) at his Costs
erect and build a good & strong wall of 12 feet in height between the sd marsh and the field
called Kae Morva Carw to prevent the Cinders Slaggs & Rubbish proceeding from the
sd works to fall into the said marsh or doing any damage thereto but if a Grant
or Lease might be obtained That then the Thos. Coster Jos. Percival Saml.
Percival and Hy Barne should have the liberty of laying & throwing the
sd Slaggs Cinders & Rubbish thereon
That the said Tho Coster Jos percival Sam Percival H Barne shod during sd term have
the use of all waters running from Melyny Vrane or elsewhere over the lands of the [18]
sd B Mansel the sd B Mansel his heirs or assigns first using it for an Engine
(if wanted) intended to be built by him or them for drawing the water from
his Coal works which sd Engine was agreed to be built in such place and
in such a manner as shod be most convenient for the water to run from
thence to White Rock afsd or where the sd T Coster Jos Percival Saml Percival
and H Barne [should think fit] to erect a Battery Mill or trial Hammer for proving of
Copper and if the water which then was or Co(uld) be brought to White Rock
afsd should prove insufficient to supply a Battery Mill or Trial Hammer
for proving of Copper that then it should be lawful for them to have
& enjoy the mill of the sd B. Mansel called the New Mill in the parish
of Lansamlet afsd the remainder of the term rent free and the same to
pull down and in the stead thereof to erect & build a Battery Mill or
Trial Hammer for proving of Copper as aforesaid if they should think
fit and in such case the sd B. Mansel should pay or allow unto them
the moiety or one half part of the expences and charges in maintaining
and keeping a man and Horse ([the word 'there' is crossed out, possibly some difficulty in reading the Lease?] to attend and carry goods to and
from the said mill if the whole expence ther(eof) shod not am(ount) to or exceed

Page 10
£20 yearly but in case the expense and charges of keeping and maintaining such
man and horse shod amount to or exceed the sum of £20 by the year that then and
in such case the said B Mansel his heirs and assigns should allow £10 yearly

17. The first recognition that slag and waste would need to be tipped somewhere. At this time the Mansels were avoiding conflict with the Duke of Beaufort who owned Morfa Carw to the north. The meadow (Cae Morfa Carw) was walled off and used as the tip whilst the main Morfa Carw field was used to build the Middle Bank works.
18. A clause recognising both the importance of waterpower and the possibility that the souces on Kilvey could dry up.

and no more such further expense to be born at the ### costs of the said
T Coster Jos Perceval Samuel Perceval and H barne without expecting any
further or greater allowance from the said B Mansel his heirs or assigns
That in case the said T Coster Jos. Percival Saml. Percival and H
Barne shod be desirous to erect and build any other mill or mills or other
buildings for their use on any other lands in the said parish of Lansamlet
belonging to the said B Mansel other and besides the premises thereinbefore
granted and demised which they shod judge most Convenient for their use
and purpose he the said B Mansel wo(uld) at the request and costs of the
said T Coster Jos. percival Samuel Percival and H Barne execute unto them
a sufficient demise and grant of the same lands so requested and all streams
waters & privileges appertaining thereto for the remainder of the said term of
51 years They the said T Coster Jos. percival Samuel Percival and H.
Barne paying unto the said B Mansel a reasonable Yearly rent for such
Lands and on which the said Mills and buildings should be erected and
not less than the then present tenants of such lands did then yield and
pay for the same.
That the said B Mansel his h(eirs) and Ass(igns) on notice to him given [19]
should during the said term of 51 years well and sufficiently supply and
furnish the said T Coster Jos. Percival Saml. percival and H Barne with
sufficient quantity of charcoal for their own use to be from time to time delivered
at the said works at the rate or price of 30 [shillings] by the doz each dozen to
contain 10 bags of char Coal and two of Braices according to the usual
measure but if the said coals to be delivered shod be only Braices
then the said Braices to be delivered at the said works at the rate or
price of 20 [shillings] the doz and if the common rate or price of char coal
Braices sold to other persons should at any time during the said term
fall or be under the rates and prices aforesaid that then and whilst the
same should be under the said rates the said B Mansel his heirs and
assigns shod abate in the said prices proportionally according to the rates
then going
Agreement between the said parties that in any case any difference or
dispute shod arise betw(een) them touching or concerning the goodness or qualities

Page 11
of such Coals so as aforesaid to be delivered at the said intended
workhouses the same should be settled and determined by 3 or more
persons skilled in coal and copper works to be nominated by both parties
and in case such persons or the majority of them shod agree and be of
opinion that the coals of the said B Mansel shod not be fit for the
purposes aforesaid or in case the said B Mansel should not be fit for
the purposes aforesaid or in case the said B Mansel his he(irs) or ass(igns)
shod desist from working of coals by reason of failure or that the same
co(uld) not be wrought to profits or in default of delivering the same at

19. Further recognition of the importance of charcoal in the copper smelting process. With provision for adjudication if the quality should vary.

the said workhouses that then and in either of the said cases but
not otherwise and during the continuance of such impediment or default
only the said T Coster Jos. Percival Saml Percival and H Barne shod
be at liberty and might supply themselves and buy Coals for the use of
the said intended works from any other person or persons elsewhere.
That an account of all such Coals that shod be delivered at the said
works for the use thereof shod be made up weekly and the money arising
due for the same after the several rates or prices aforementioned should
be paid unto the said B Mansel his heirs and ass(igns) at the end and
expiration of every three months after the same shod become due all such
coals to be measured by a person to be jointly nominated by the said
parties to which sd person ye said Thos. Coster Jos. Percival Samuel
Percival & H Barne shod pay 4d for every day he should be employee
in measuring of the said Coals

That the said T Coster Jos. Percival Saml. Percival and H Barne
shod on reasonable notice (if they or their Agent for the time being shod
require the same) have always a stock of coals delivered at or near the sd
workhouses and not less in quantity than 20 weighs

Covenant by said T Coster Jos. Percival Saml. Percival & H Barne
that they would not during the said term buy or make use of any Coals
in the said intended workhouses Mills or other the premises aforesaid other
than the said Coals of the said B Mansel his heire and assigns as
long as he or they could supply them therewith and shod deliver the
same as aforesaid and so that the same coals so to be delivered by the said
B Mansel shod be good clean and marketable merchantable and fit for the several
uses of the said intended workhouses

That the said B Mansel shod supply sufficient quantities thereof at the

Page 12
said workhouses as shod be from time to time by the said T Coster Jos.
Percival Saml. Percival and H Barne reasonably required or demised
That the said T Coster Jos. Percival Saml. Percival and H Barne
shod pay unto the said B Mansel the yearly rent before reserved clear of
all taxes whats(oever)
And shod at the end of every 3 months pay unto the said B Mansel
21 [shillings] by the weigh for all Coals to be delivered at the said works according
to the measures aforesaid and also 30 [shillings] by the doz for every doz Bags of
Charcoal and 20 [shillings] for every dozen Bags of braises and 35 [shillings] by the ton
for every ton of timber
That in case default should at any time be made in any or either
of the several payments agreed to be made the said B Mansel his heirs
or assigns shod have power to distrain for the same.
That the said T Coster Jos. Percival Saml. Percival and H Barne
shod at all times during the said term oblige all such Masters or owners
of Ships and vessels that should at any time being any oar or other
materials to the said works to load all such Coals as they should Ship
or take into such vessels during such respective voyages at the said Coal place

or Bank called White Rock aforesaid at the usual rates or allowances
the same were sold at by the said B Mansel his heirs or ass(igns) to o(ther)
persons in case there should be sufficient to load the same
Proviso for distress and entry on nonpayment of rent or breach of
covenants
Covenant by the said B. Mansel for peaceable enjoyment
Executed by the said B. Mansel and attested
by 2 witnesses

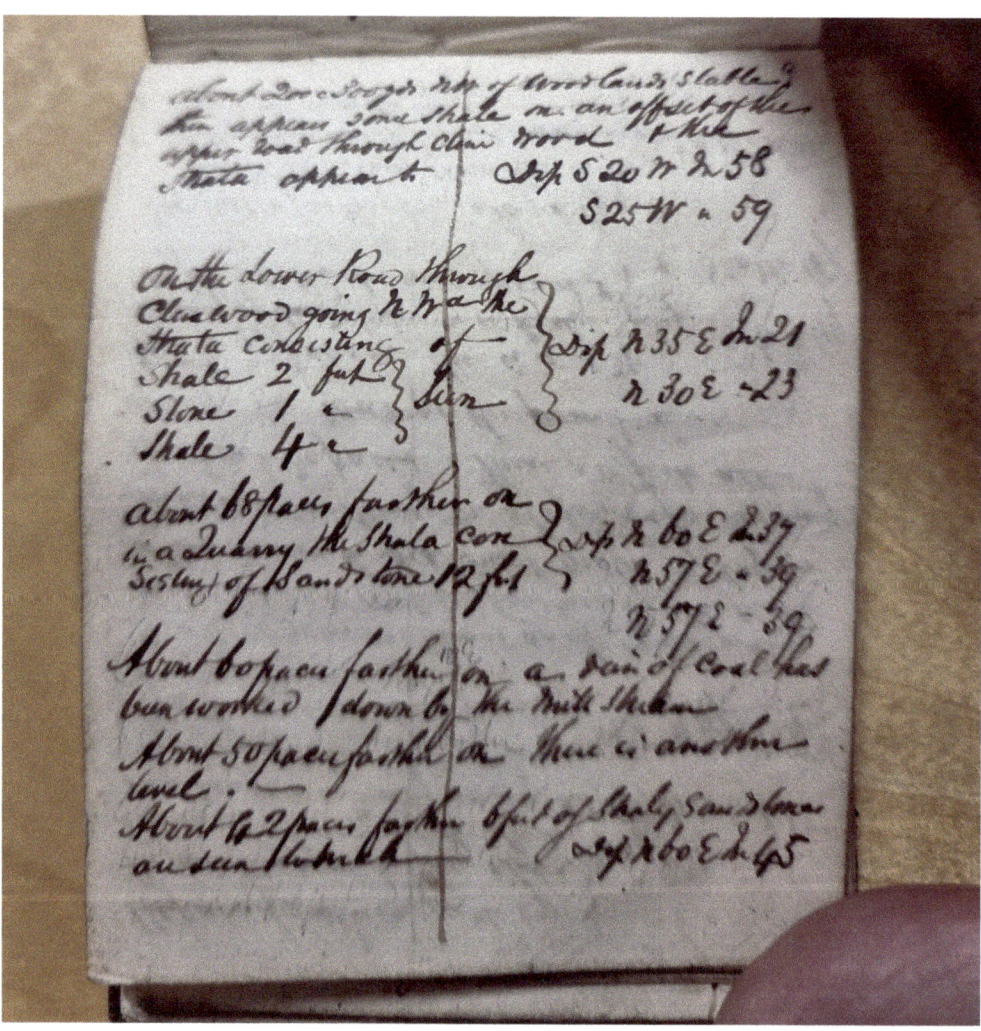

Above: A page from one of Logan's small notebooks. Logan would usually write in pencil when working and transcribe into ink later.

Annex 2: Extracts from William Logan's Fieldnotes 1837-40

As is normal with a field geologist, Logan kept a pocket notebook and used it extensively on his many fieldwork tours of the area. Before moving to Canada in 1842, Logan donated his files and notebooks to the British Geological Survey. The notes were heavily used by the later geologists that continued Logan's work on the geology of the South Wales Coalfield. The primary guide to the Swansea area (the Memoir, which is referenced in the chapter on Coal and Kilvey), which was published in 1907, also relies heavily on Logan's original explorations. Although meant as a technical geological guide to Swansea, the Memoir has now become a historical document itself as it describes in considerable detail the landscape of Swansea before twentieth-century demolition and development destroyed so much.

Logan's notebooks are mainly small (8.5cm by 11cm) notebooks, made of good quality paper and with a small brass clasp to keep them closed and protect the writing from rain and debris. Upon opening one in January 2023, I saw small amounts of fine rock dust drop onto my own notebook from the spine, just as my own geology notebooks have from being carried around in a pocket with rock samples!

Normally, Logan would write in pencil in the field and rewrite in ink afterwards. His small series notebooks were not normally dated but archivists in the BGS Archive have determined the sequence in which they were written. This suggests that logan visited Kilvey early in his fieldwork in 1837 and came back many times later as he worked on the stratigraphy of the coal veins and investigated the extents of the various faults in the north of Kilvey. When Logan started working with Henry de La Beche and the BGS staff he transcribed all of his notes into a large format notebook with each location visited in alphabetical order.

Logan would frequently talk to mine workers, engineers' surveyors and colliers and was always on the hunt for the lists of sequences of rocks and coal veins seen by these men as they dug adits and pits. Some local colliers kept extensive records of the nature of the rock layers they were digging through and Logan's notes have many such stratigraphies with depth or thickness information in fathoms, feet and inches. Many of the coal veins and localities mentioned here are shown on the sketch map on page 21.

1. The pocket notebooks
Book 1 (has a pencil date of 1737 entered by an archivist)
Page 21
The Benson + Logan Co. had another
pit 12 fm deep where the
bridge to Forest Uchaf was
Near Forest Uchaf Calland
bored 100fm through cliff
+ stone but found no coal
at Pont y Blawdd Sir J. Morris

bored 80 fms + came to the
Church Vein
Morgan – bored 60 fms above
Drymau isaf

Between the Church + Middle
vein there is a small vein
of 20 inches + between the
Middle + Great Vein there
is a small vein of 9 inches

say 22 fms below Middle Vein

The vein on the West of
the Engine Vawr fault near
Bonymaen is the 2 feet Vein
+ that cropping out by White [1]
Rock is Hughes Vein. Near the
Brook it crops out + a little
above enters the Hill again
seen on the very top of it

Here in Section
[small drawing, see page 25]
It dips 4½ feet in a fathom
There is a level on it at
White Rock Works + another
on the Rotten Vein which
12 fms above. The one level
is on one side of the Brook +
the other on the other

Page 50
[from] C.H. Smith
The following veins have been
cut through by a level on the
full rise of the Hill behind
White Rock

			ft	in
Warky Vein	Coal		1	2
	Rub(bish)		4	
	Coal		-	9

36 yards from which
in a Horizontal line
on the full rise

Small Vein	Coal	1	8

22 yds from which

Rotten Vein	Coal	3	6

22 yds from which

Hughes Vein	Coal	5	-

In Cilfay Hill below Hughes
Vein is the Captains Vein
below that Foxhole Vein the
Crop of which is seen on the
Westside of the Hill above Foxhole
where it was – by C H Smith
+ found to be 18 inches and
again on top of the Hill
also – by C H Smith where it
was found to be 5 feet

Book 4
Page 37
Cilfay Hill 22 April 1839
Almost due South from Burley Hill
House to the South of the Brook running
to Ty Coch there is a considerable
Spring + to the Southward of
Pen Y Graig Uchaf on a line with
Ty Draw there are 2 more + others
a little farther up the Hill
Between the First mentioned spring
+ Cwm Bach there is another +
along the whole of the eastern
side of the Hill about the Place
is a apparently a line of them
There cannot be much doubt
these are on the great Bony
Maen fault.
25 yards south of the Carn on
Cilfay Hill is the top of the
Seam of coal which Smith
drove a level on above Foxhole
Above this seam of coal there
is 150 ft horizontal of

---

1. Logan spent a lot of time working out the relationships of the various coal seams (veins) across the Hill. The sketch map on page 21 shows his eventual conclusions. The constant challenge was trying to distinguish between the Hughes and Foxhole veins, and you will see these mentioned frequently.

sandstone say about 100 feet
thick – seen on the road
over the Hill –
Below the same seam are
seen in the quarry on the side
of the Hole above Foxhole about
120 feet thickness of sandstone –
Near the Vein is a thin seam of
shale say 10 inch thick =
has an undercliff [2]–

Page 38

Cilfay Hill 23 April 1839

[This page was in pencil and Logan has inked over it. Some of the legible pencil is different to his ink sentences. it looks to have been a long day of work with a complex section taken]

From the pits on Hughes to the
south of White Rock Brook
going north-
At 115 ft there is a line of pits on a vein
233 ft is a line of pits + sandstone
Rock immediately above
50 feet thick
350 ft is a pit to Coal
At a less distance than 50 yards to south
of the Pits is Hughes Vein Sandstone
Measuring Southward and from the Bridge
- over White Rock Brook for
Rail Road
At 68 yards Sandstone appears of
which the follow is exposed on the
Rail Road
Hard Grey Sandstone thick beds 37 yds
Coarse sandstone with Iron Stone

- + - collection
of - + upland to Ironstone 2 yds
Sandstone Hard Grey 23 yds
- Coarse –
- + - - stone 2 yds
Do. Hard Grey thick beds 13 (ft)
Coal a - 1 (ft)
Sandstone coarse 1 foot
Shale 2 foot
[stratigraphically] under the above but seen higher
up the side of the Hill
Sandstone within laminae
little argillaceous 20ft
Sandstone thin splitting 45ft
Do. Coarse full of casts of
Stony – replaced
by Iron stone 2ft

Sandstone	20ft
Sandstone thin splitting	30
Coal supposed about	6in
Clay with - -	2ft
Sandstone	8
Shale bluish	5
Sandstone and shale	6
Sandstone thick bedded	35
Sandstone + beds very regular + very hard excellent building stone micaceous	34
Sandstone – regular Beds	8

Page 39

Sandstone micaceous, enclosed
nodules of ironstone –
pebbly dimensions + some

2. Logan uses the word 'cliff' or 'clift' to describe what we would later call seat earths or clays. These words were local terms for the clay, it could also be called 'Cornish clay'. These clays had a very close association with coal seams were an important part of interpteting the size and position of underground coal seams. Many cliffs provided good clay for bricks as can be seen in the remaining structures of the White Rock works.

6 inches in diameter		
Shale		2
Sandstone thinly laminated + shale		
Sandstone probably with perhaps a band of Shale in the middle of it		150ft
Coal a seam opened by Smith above foxhole + by him found to be useful for smelting 6ft + some- only 18 inches		2ft
Sandstone below this		20ft

[In pencil]
In Tany Graig quarry
about 90 feet are exposed
The lower 70 sandstone
Dip S2W43
some shale

[In ink]
In the quarry above Tanygraig
about 90 feet are exposed
In the 20 feet at the top there
are some small beds of
Shale The lower 70 feet
is Sandstone
Dip S2W43.

Page 40 [3]

	ft	ins
Coal Warky Vein 4ft Rubbish	5	
Underclay	3	
Sandstone	50	

Shale		16	
Coal Small Vein		1	8
Underclay		2	
Shale		40	
Rotten Vein		3	6
Underclay		3	
Shale		40	
Coal Hughes Vein		5	
Underclay	130 172.2	3	
Sandstone		17	5
Coal		0	6
Underclay		2	
Sandstone		288	
Coal		10	
Underclay	540 467.4	1	0
Sandstone		8	
Shale		2	
Sandstone		167	
Shale		2	
Sandstone		160	
Shale		3	
Sandstone		160	
Coal		2	
Underclay	350 449.6	2	
Sandstone		200	
Coal a small seam		6	
Underclay		1	
Shale		110	
Sandstone		100	
[total]		1500	5

3. This stratigraphic sequence is one of Logan's early attempts to sequence the layers of coal and sandstone in the Pennant Sandstones Formation on Kilvey. Logan is already recording the closé relationship between Coal and Underclay which formed the scientific basis for understanding the methods of formation of coal seams in Carboniferous rocks. Coal veins were almost always situated on a bed of cliff or underclay. Veteran colliers knew this and it helped them search for coal veins in underground explorations.

2. The Large Notebook

The Llansamlet entry covers pages 29 to 38 of the original large notebook. These pages also have a series of pencil thickness calculations and summations over the original ink entries. I have omitted these from here to avoid confusion. The pencil figures suggest Logan was calculating coal vein and rock layer thicknesses for a full stratigraphy of the Swansea area. The pages transcribed here cover White Rock and western Kilvey, pages 36 to 38 cover Llansamlet and further north. Logan's ability with the Welsh language and placenames was superior to most of his colleagues and he quickly learned his way around the hills of Kilvey and Townhill.

These notes show a number of things happening, Logan is teaching himself geology, he is talking to local experts and he is developing his own theories of how the western coalfield was constructed.

Page 29

Llansamlet

Various information respecting the Coal Veins +

measures in

the taking of C. H. Smith Llansamlet

Section of the ground at the Doublepit above Tyr Back

from actual cutting by C H Smith

	Fms	Ft	
Sand + gravel	2	3	
Mettle Vein Rock Very Hard	7	1	
Cornish Cliff	3	3 [1]	
Coal Middle Vein		5	
Undercliff	1 [2]		
Soft Cliff	9		
Mixed Stone + Cliff	8	4	
Soft Cliff	17	2	6

Big Rock (Very Hard)	4	5	6
Soft Cliff	12		
Coal Little or Jenkin Vein		1	
Undercliff	1	2	
Mixed Rock + Cliff not much rock	18		
Soft Black Stuff (Cornish)	1	3	
Coal Great Vein		1	
Undercliff		5	
Black Soft Stuff			3
Strong Rock	4	3	
Cliff	1	1	
Coal Three ft Vein		3	

Page 30

Section of the Ground at the Great Pit to the east of the Fault from the Middle Vein down wards C H Smith

Middle Vein [Probably the 5ft Vein]

50 fms

Small Veins

- No. 1 4ins
- No. 2 3ins
- No. 3 9 ins

9 fms

Small Vein   15ins

10fms

Great Vein   60ins

          8fms

Three feet Vein   36ins [Probably the 6ft Vein]

          20fms

Two feet Vein   24ins

As the above pit was sunk near the great fault which throws the Measures down to the Westward at Bonymaen it is probably in the upper part affected by it

Page 31

Section of the Ground + Veins of Coal at Charles Pit near Llansamlet Church according to Evan Benjamin

Surface

          25fms

Church Vein   with Cliff   4ft [4Ft Vein]

          4fms

Small Vein   2ft

consisting of Coal 1

     + Dirt   1

          61 fms

Middle Vein   6ft

It has a stone roof

Consists of Rub   2ft

Coal 1ft 6ins

Shale 6ins

Coal 2ft

    30 fms

Small or Jenkin Vein   2ft

    It has Cliff for Roof

    Consists of Coal 1ft

        Dirt 1ft

        30 fms

Great Vein   6ft

    It has for Roof some –

        Stone and Stone – Cliff

        Consists of Coal 4ft

Dirt 1 ft

Coal 1ft

7 fms

Three feet Vein   3ft

    It has Cliff for Roof

The Great fault at Lansamlet throws the measures up to the East 40 fathoms [The Park Pit Fault]

The – pit to the South of Dyffryn – is 11 fathoms + the Middle Vein is 300 yards to the South of it. The pit is sunk in the fault.

Page 32

On an old map of the workings of C Townsend in the possession of C H Smith of Gwernllwynchwyth, among various remarks on the Veins of Coal are the following.

A Level was driven from the Middlebank Copper Works to the Warky [4]

Vein + along it 700 yards to the east

> The Warky Vein
>
> Consists of   Coal 0ft 9ins
>
> Rubbish 4ft
>
> coal 1ft 2ins

A level was driven across the measures from the Warky Vein 36 yds

> A Small Vein
>
> Consists of Coal 1ft 8ins

which was worked along the level course up the dingle 64 yards

A Level was driven from the 20 Inch Vein across the Measures

22 yards to

> The Rotten Vein
>
> consists of Coal 3ft 6ins

The Vein was worked along the level course 66 yards to the east

The coal was bad and so was the roof, requiring double timbers

all the way. A level was driven from the Rotten Vein across

the measures 22 yards to

> Hughes or Foxhole Vein
>
> Consists of Coal 5ft

A level was driven from the Upper Bank Copper works to

> The Twofeet or Cwm Vein
>
> Consts of Coal 2ft

4. See the sketch map on page 21.

At the bottom of a pit on the Vein a hole was bored to the depth of 49½ fathoms. The only Seam of Coal met with in the bore hole was at 8 fathoms

  A Small Vein

Consts of Coal 1ft

Page 33

Information from C H Smith + Mills respecting the Coal at Lansamlet [5]

The Level course of the coal at the bottom of the Charles Pit near lansamlet Church in the middle vein 50W+S50E. The dip of the Vein is 1 in 6. The level course by which the Colliery is worked is about 10 fathoms above the water level Course and it meets the Cwm fault about 20 yards in front of the door of the School House. The Cwm fault throws the coal up 53 fathoms to the east. Smith has continued the Level through the fault, which is plainly seen first throwing the coal up about 10 feet + then the remainder

of the distance about 40 to 50 yard further on. the

  Small or Jenkin Vein

which is 22 fathoms above the great vein is seen on the

5. The faults at Cwm and Llansamlet proved difficult to understand and were very disruptive to the coal seams underground. The faults here were frequently filled with broken coal and sandstone and caused a disruption of over 70m (240 feet) in the coal veins across the faults.

east side of the fault + runs along the level 300 yards

The real width of the fault is in this cutting. Shewn

to be but a few feet wide or rather a few inches. A

leader of coal runs up it + dips to the west about 75°

After running along the Strike of the measures to about

300 yards east of the fault, the drift turns half course South

for 50 yards + then to the full rise of the measures

towards the Great Vein

   The Middle Vein [Swansea 5 feet] [6]

   Consists of Roof Rock

     Coal 3ins

     Cliff 8ins

     Coal 3ins

     Cliff 4 ins {called Rubbish}

     Coal 2ft

     Cliff 3ins

     Coal 3ft

     Cliff 3ins

     Coal 6ins

     Undercliff

Page 34

The Church pit near Gwernllwynchwyth is 80 fathoms deep to

6. You can see here again the relationship between coal veins and 'cliff' or clay.

The Church Vein

The Vein has been worked 280 yards down to the -. the dip at the bottom of the pit is 4 inches [per] yard but it lessens to the - . the lowest level must be pretty near the centre of the basin or saddle. Going eastward it gradually turns northward,- - -

- throw fault to Mynydd Drymma. At this point the LC runs nearly at right angles to its original direction which at the bottom of the pit is about n45W + S45E. About yds to the east of the pit there is a down throw to the east of about 10 fathoms

The Roof is Cliff

The coal is 4ft thick

There is a thin layer of shale in the middle. To the - the Coal is very hard. It is necessary to blast it with gunpowder.

According to Mr Elias Jenkins there are three veins of Coal under Cilfay within workable reach. They are

Hughes Vein

Captains Vein

Foxhole Vein [7]

On the west side of the Hill above Foxhole Smith drove

---

7. The Hughes and Foxhole veins were easily confused and probably had been for centuries. the Hughes was always reliable but the Foxhole tended to vary in thickness making it a risk to mine. There are several scars remaining on the Hill above White Rock and Foxhole where early mine workers 'chased' the vein up the Hill See the image on page 29.

a level into a vein which he found to be 1ft 6in. On the top of the Hill he tried it again a found it there to be 5ft thick, and it is consequently considered to be a lumpy vein.

Page 35

Information obtained from Evan Daniell of Ty Gwyn formerly agent to [left blank] on the Llansamlet Coal district.

The perpendicular distances between the chief seams are

From Church Vein to Middle Vein	55fms
Middle to Great	77fms
Great to Three feet	8fms
Three feet to Two feet	18fms
Two feet to Rotten	108fms
Rotten to Hughes	12fms

The level course of the Charles and Church pit seams is from N45W + S45E

Between the Engine Fawr fault + the great Llansamlet fault there are two others. The one near the Engine Fawr throws the coal 15 fathoms up to the east. The other throws it up to east also, but only a few yards

Between the two great faults the Vein dips 1 in 10.

At the air pit near Dyffryn aur, which is on the edge of the great fault, the Cwm level is lower than the Coal on the east side of the fault. The level + the Coal cross one another about 50 yards farther back in the level + a

proportionate number down the Slope of the Vein. The
Level course of the Coal from the point of crossing went to
about 10 or 12 yards north of the Summer House.

There is a joint to East of the Strait road to Trallwyn
throwing the Coal up 5 feet

From Engine Fawr there is a horse way down to the great Vein
near the halfway house 1300 yards.

Between the Church + Middle Vein there is a small Vein of
20 inches + between the Middle + great Vein there is a small
Vein of 9 inches. It is 22 fathoms below the Middle Vein

The vein worked on the west of the Engine fawr fault near
Bonymaen is the 2 feet vein + that cropping out near White
Rock is Hughes Vein. The lumpy vein on the top of Cilfay hill is so

Page 36

like Hughes Vein, that this, after cropping out by White [8]
Rock was supposed to enter the hill again + appear up
at the top. The Stratification however shews this cannot
be the Case, unless there were an east + west fault.

Hughes Vein dips 4½ feet per fathom. There is a level on it
at White Rock works + another on the Rotten Vein. The
one is on the one side of the Brook + the other on the other.

A level was driven into Cilfay hill near Cwmbach
to try for Hughes Vein, but it was not found, neither was the

8. Here Logan tries to settle the age-old confusion between the Hughes and the Foxhole veins by using stratigraphy to identify the Hughes, Rotten and Foxhole veins. It would be revised again in the 1890s by Aubrey Strahan, although Logan was commendably accurate.

Rotten Vein.

Information on the Llansamlet Coal from Mr David now of Penclawdd, who formerly worked in Smith's Colliery There is a level from the Engine Fawr to the Halfway House to _ through the Great Fault. It is 1700 yards long The coal on the west side of the fault in the Middle Vein is 3 feet + has a layer of Lymed [sic] under it. On the East side of the fault it is 2ft 9ins + has under it a layer of stone instead of Lymen [sic]. On the West side of the fault the Church Vein is 4 feet on the east it is 2 feet 3ins. The Great Vein on the east side is 6 feet +on the West 4ft 6 ins

In the course of the 1700 yards above mentioned there are Two down throws. One of 6 feet + the other of 9 feet Old Martin used to say the Llansamlet Veins were to be found over at Aberavon. [9]

---

[9]. 'Old Martin' is Edward Martin, see page 24. Martin's notes from about 1806 were still providing basic coalfield guidance in the 1830s!

Index

Symbols

1306 Charter 17

A

Aberdare 89
Afon Llan 14
  Llan Valley 22
Ancient Woodland 88
Andrew's Bakery 13

B

barometer 31
Barry Docks 68
Benson and Logan Company 30
Berlin 82, 89
Birchgrove 20
bombsite flora 89
British Association 32
British Geological Survey 31
British urban natural history 70
  British urban habitats 82
Burley Hill 31

C

CADW 37
Cae Morfa Carw 40, 48, 54
Calland coal pit 30
Calland Pit 105
Cambrian, newspaper 52
  Vivians' Copper Works...as a Public Nuisance' 1833 52
Carreg Cennen Castle 32
chalybeate springs 17
Charcoal 102
Cilfái 13
Cilfay 30
climate change 20
Cnap Goch House 56
Coal 17
  Coal and undercliff 107
  coal seam cyclothem 31
  coal seam faulting 32
  early coal mining outputs 18
  early Kilvey coal mining 18, 48
  Mansel Coal Yard at White Rock Quay 57
  White Rock Wharf 38
Coal Veins
  Captain's vein 30
  Church 30, 105
  Dyfatty vein 14
  fireclay at the base of coal veins 31
  Foxhole vein 14, 18, 80, 106
  Great vein 105
  Hughes vein 18, 30, 80, 106
  Penlan vein 14
  Popkin's 100
  Rotten vein 14
  the 'dip' of a coal vein 24
  Warky vein 14
common badger, 17
Copperopolis 11
Copper ore
  chemical composition of the Cornish copper ore 48
Copper slag
  Tipping 47
    Cae Morfa Carw 44
Copper Smoke Trials
  Copper Smoke 62
  copper smoke trials of the 1830s 52
  impacts of smoke on vegetation and animals. 52
  witness descriptions of the copper smoke 54
Cornish copper ore 48
Coster, Thomas 40
Coster, Thomas, Lessee 1737 93
Cwmafan 71
  Copper Mountain 71
Cwmbach 31

D

de Breos, William 17
decolonisation movement 80
De La Beche, Henry 26, 32
Drift Mining
  drift mine, 18

E

Ecosystem Services 86
Enterprise Zone 56

F

Fabian's Bay. 13
first edition of the geological survey map for Swansea 32
first industrial nation 20
Forestry Commission 64, 86
  an impressive forestry portfolio 70
    ploughing 66
Forest works 31
Foxhole 18, 30, 80
  Foxhole coal shipping quays 18
Francis, George Grant 47, 48

G

Geology
  customary and oral tradition 28
  Engine Vawr Fault 105
  fireclay 31
  geological sections 32
  geology 22
  glacial gravel 57
  Pennant Sandstone 22, 57
Gilbert, O. L. 82
Glamorganshire Sheet 37 (Glam. XXXVII) 26
Great Bon-y-maen Fault 31
Great Western Railway 68

H

hachures 28
Hafod 11
  Hafod works 47, 50
  Pentre Hafod 50
Hafod chimneys 14
Hilton, K.J., 61
Hinchliffe, Steve 84

I

inclined planes 50
inclined tramway 56
industrial archaeology 72
industrial nature 88

J

Japanese Knotweed 89
Jenkins, Elias 30

K

Kilvey
  a distinctive post-industrial ecological community 89
  beneficial outcomes of repair and restoration 80
  complicated leasehold and freehold land ownership 64
  described as 'barren as a road 54
  ecological surveys 86
  environmental condition has gradually recovered 80
  Kilvey Hill and its regeneration as part of of the wider area 72
  lost its protective status within Welsh national forestry 84
  transition from industrial forest to community woodland 71
  woodland area becomes Plot 57 64
    commercial planting of the hill 70
    Plot 57 becomes Kilvey Woods 71
Knap Goch (or Knaploth) 13
Knaploth Mill 38, 40, 94
  millpond 42, 50
  sluice 44
  the corn mill originally above the White Rock site 78
Kowarik, Ingo 89

L

Lachmund, Jens 82
  ecologists were often regarded as 'unreliable friends 84
  transition from conservation ecology to urban nature conservation 82
Landore 11
Latour, Bruno 82
legacy landscape 47
Le Play, Frederick 48
Llan, River 57
  Cadle 57
  Glyn Silling 57
Llansamlet Church 31
Llwynheiernin 24
  Llwynheiernin Farm 40
local colliers 28
local volunteer groups 78
Logan, William Edward 24
  coal vein thickness and dip 28
  conversation with a local collier Elias Jenkins 30

early studies 26
exploring the geology of coal 28
hand lens, compass, and clinometer. 31
Kilvey section 32
Logan had several advantages 28
the first edition of the geological survey map for Swansea 32
Upper Forest 26
walking around Morris Lane, Foxhole 30
Lordship of Gower 18
Lower Swansea Valley industries 9
Lower Swansea Valley Project 61, 78
early 1960s revegetation research 66
Luftwaffe, 1941 61

M

Malm, Andreas 20
Mansel, Bussy, fourth Baron 93
Mansel Estate 38
  Mansel Coal Yard 57
  Mansel Estate 38, 40
  Mansel family in Glamorgan 48
Margam 68
  Margam Forest 68
    The 'great fires' of March 1929 68
Martin, Edward 24
Massey, Doreen 80
  trajectories 80
Merchant, Carolyn 47
Middle Bank 14, 32
  Middle Bank 61
Morfa Carw 40
  Cae Morfa Carw 44
Morganite 56
Morris Lane, Foxhole 30
Morris, Sir John 30
Morriston 28
Mount Pleasant 13

N

Nant Llwynheiernin 14, 31, 44, 66, 77
  original features that Logan surveyed 32
Neath-Swansea mail coach 54
non-native plants 88

O

Ordnance Survey 26
Orienteering 71
Outlines of Geology 26

P

Penllergaer 14
Penllergare Estate 22
Pen y Graig Uchaf 31
'pioneer' trees 10
Plot 57. See Kilvey: woodland area becomes Plot 57
Pont y Blawdd 105
Port Talbot 68
Port Tennant 32
Posi-industrial woodlands in Wales 70
  changing attitudes 71
  multi-purpose' woodlands 71
post-industrial recovery 9
Post-industrial woodlands in Wales
  a distinctive post-industrial ecological community 89
Prevailing winds 48
Pugsley, David 89

R

Rio Tinto 71
Royal Navy 38

S

Seyler, Clarence 17
Slave trade 80
Smelting
  Welsh Method' of copper smelting 48
Smiths Canal 44
Smith, William 18
spontaneous' vegetation 88
Steam engines 47
  1805 lease 50
  steam power 52
Stopes, Marie 17
Sukopp, Herbert 82
  urban nature around the bombsites, 82
Swansea
  Glamorganshire Sheet 37 (Glam. XXXVII) 26
  landscape in 1830s Swansea 28
Swansea Council Estates Department 64
Swansea Council Parks Department 66
Swansea Scientific and Field Naturalists Society 70
  Swansea's bombsites vegetation communities 89
Swansea's early industrial revolution 61
Swansea valley copper 38

T

Tawe, River

    attempts to prevent tipping in the river 57
    Morfa Quay 56
    navigation of the river 50
    Town Reach and Fabian's Bay 50
Teagle, Bunny 70
theodolite 28
    theodolite 31, 32
Tir-isaf 20
Town Hil 30
Townhill 13
Trees
    Douglas Fir 68
    first 200 'gasworks' trees 66
    Larch 68
    Lodgepole Pine 68
    Monterey Pine 66
    oak 89
    old trees and the dead wood 88
    Scots Pine 66, 68
    Sycamore 99
Turnpike Road 44
Ty Coch 31
Ty Draw 31

U

'undercliff' or 'cliff'. See also fireclay
University College of Swansea 62
Upper Forest Works 24
urban nature 86

V

Viking settlement 78

W

water power engineering 20
    extensive leat network 77
    waterpower 50
    waterpower resource from Kilvey Hill 40
Welsh Development Agency 62
Welsh Post-industrial Woodland 10
Werner, Abraham Gottlob 18
White Rock 13
    Agreement For Sale, 24 August 1735 38
    copper furnaces 47
    devolved into an industrial oven 62
    location 47
    Morfa Carw 40
    Morfa Carw meadows 57
    White Rock, early coal 18

White Rock now lifeless 47
White Rock Quay 57
White Rock ruins 61
White Rock site, as leased in 1735 56, 94
White Rock Brook 31
  White Rock Brook 31

White Rock Brook or 'the Dingle' 40
White Rock Heritage Park 50, 57
Witham, Henry 17

www.ingramcontent.com/pod-product-compliance
Lightning Source LLC
Chambersburg PA
CBHW060927170426
43193CB00022B/2981